MATERIALS SCIENCE AND TECHNOLOGIES

A CLOSER LOOK
AT PYROPHOSPHATES

MATERIALS SCIENCE AND TECHNOLOGIES

Additional books and e-books in this series can be found on Nova's website under the Series tab.

MATERIALS SCIENCE AND TECHNOLOGIES

A CLOSER LOOK AT PYROPHOSPHATES

RITESH L. KOHALE
AND
SANJAY J. DHOBLE
EDITORS

Copyright © 2020 by Nova Science Publishers, Inc.

All rights reserved. No part of this book may be reproduced, stored in a retrieval system or transmitted in any form or by any means: electronic, electrostatic, magnetic, tape, mechanical photocopying, recording or otherwise without the written permission of the Publisher.

We have partnered with Copyright Clearance Center to make it easy for you to obtain permissions to reuse content from this publication. Simply navigate to this publication's page on Nova's website and locate the "Get Permission" button below the title description. This button is linked directly to the title's permission page on copyright.com. Alternatively, you can visit copyright.com and search by title, ISBN, or ISSN.

For further questions about using the service on copyright.com, please contact:
Copyright Clearance Center
Phone: +1-(978) 750-8400   Fax: +1-(978) 750-4470   E-mail: info@copyright.com

NOTICE TO THE READER

The Publisher has taken reasonable care in the preparation of this book, but makes no expressed or implied warranty of any kind and assumes no responsibility for any errors or omissions. No liability is assumed for incidental or consequential damages in connection with or arising out of information contained in this book. The Publisher shall not be liable for any special, consequential, or exemplary damages resulting, in whole or in part, from the readers' use of, or reliance upon, this material. Any parts of this book based on government reports are so indicated and copyright is claimed for those parts to the extent applicable to compilations of such works.

Independent verification should be sought for any data, advice or recommendations contained in this book. In addition, no responsibility is assumed by the Publisher for any injury and/or damage to persons or property arising from any methods, products, instructions, ideas or otherwise contained in this publication.

This publication is designed to provide accurate and authoritative information with regard to the subject matter covered herein. It is sold with the clear understanding that the Publisher is not engaged in rendering legal or any other professional services. If legal or any other expert assistance is required, the services of a competent person should be sought. FROM A DECLARATION OF PARTICIPANTS JOINTLY ADOPTED BY A COMMITTEE OF THE AMERICAN BAR ASSOCIATION AND A COMMITTEE OF PUBLISHERS.

Additional color graphics may be available in the e-book version of this book.

Library of Congress Cataloging-in-Publication Data

ISBN: 978-1-53617-730-5

Published by Nova Science Publishers, Inc. † New York

CONTENTS

Preface — **vii**

Chapter 1 Introduction to Pyrophosphate — **1**
Ritesh Kohale and Sanjay Dhoble

Chapter 2 Synthesis Techniques of Pyrophosphate Phosphor Materials — **21**
Dinesh Kumar and Ram Sagar Yadav

Chapter 3 Optical Properties of Lanthanide-Based Pyrophosphate Phosphor Materials — **47**
R. S. Yadav, Monika, S. J. Dhoble and S. B. Rai

Chapter 4 The Pharmacology of Pyrophosphoric Acid: A Critical Analysis of the First Quarter Century of Research — **71**
William Banks Hinshaw and Allyn F. DeLong

Chapter 5 The Analogous Causation of Phossy Jaw, Osteonecrosis of the Jaws, Induced Femur Fragility, and A typical Femur Fractures — **89**
William Banks Hinshaw and Louis DuBose Quin

About the Editors 121

Index 125

PREFACE

Present book is edited and developed in support of the researchers, students working in the field of material science and technology professionals in the field of pyrophosphates.It provides learning resources and scientific ideas for the development and implementation of the ideas.

The foundation behind the book is that readers will have ample opportunities to enrich their learning experience and extend a range of abilities through exploring the fundamental concepts about pyrophosphates along with their applications. Carefully edited, written and sequenced, the materials and content in this book aim to:

- Strengthen readers skills of understanding and ideas;
- Help readers to respond for scientific and technological performance means;
- Enable readers to apply the knowledge and skills they have learned in their own creative production and critical appreciation of study.

Present book encompasses and gives brief information about pyrophosphate materials. And it also explains their phenomenological theories, properties of rare-earth activated pyrophosphates. This book provides the comprehensive in depth information about the development of phosphates and pyrophosphates along with their essential characterization techniques as well as their diversified applications in numerous domains. Present book envelopes all essential information about single and mixed cations pyrophosphate compound that are used from past few years. It also covers intact information about innovation in advance pyrophosphate materials especially for their applications in medicinal field. In particular this book focuses on enhanced chemical and physical properties of Pyrophosphate as there is a need of new low cost and less toxic, energy efficient materials to reduce the current cost of device technologies. This book gives a masterpiece information about present scenario of pyrophosphates and concurrent technologies with its increasing significance to resolve the energy and environment related problem and their sustainable approach in the world of globalization.

In: A Closer Look at Pyrophosphates
Editors: Ritesh L. Kohale et al.
ISBN: 978-1-53617-730-5
© 2020 Nova Science Publishers, Inc.

Chapter 1

INTRODUCTION TO PYROPHOSPHATE

Ritesh Kohale[1] and Sanjay Dhoble[2]

[1]Department of Physics, Sant Gadge Maharaj Mahavidyalaya,
Hingna, Nagpur, MH, India
[2]Department of Physics, R. T. M. Nagpur University, Nagpur, India

ABSTRACT

In general, a phosphate is derived from phosphoric acid. The phosphate ion $(PO_4)^{3-}$ is an inorganic chemical that can produce numerous derivatives. In organic chemistry, a phosphate, or organophosphate, is a customized form of phosphoric acid. Amongst the several phosphates and phosphoric acids, organic phosphates are significant in geochemistry and biochemistry as well as ecology. Moreover, inorganic phosphates are extracted to acquire phosphorus for use in industry and agriculture. At prominent temperatures in the solid-state, phosphates can reduce to obtain pyrophosphates. In biology, adding phosphates to and removing them from proteins in cells are both essential in the directive of metabolic developments. Mentioned to as phosphorylation and dephosphorylation, individually, they are significant techniques that energy is deposited and loose in active systems. The science and technologies moved forward for the development of efficient phosphate based phosphors that may be applicable for mercury-free lighting solutions. There is an ever-increasing demand for novel ultraviolet (UV) and vacuum-ultraviolet (VUV) phosphates, scintillators, and laser materials, which have been wide application in numerous lighting applications. Luminescence materials doped with lanthanide ions have attracted much interest for its blue light emission. Europium doped pyrophosphate phosphors are used as a tricolor component in fluorescent lamps and phosphor-converted white LED.

INTRODUCTION

The phosphate ion is a polyatomic ion using the experimental formula $(PO_4)^{3-}$ and a molar mass of 94.97 g/mol. It contains one principal phosphorus atom enclosed by four oxygen atoms in a tetrahedral surrounding. The phosphate ion conveys a −3 recognized charge and is the conjugate base of the hydrogen phosphate ion, $H(PO_4)^{2-}$, which is the conjugate base of $H_2(PO_4)^-$, the dihydrogen phosphate ion, which in turn is the conjugate base of phosphoric acid, H_3PO_4. A phosphate salt proceduces when a positively charged ion attributes to the negatively charged oxygen atoms of the ion, developing an ionic compound. Numerous phosphates are not soluble in water at typical pressure and temperature. The sodium, potassium, rubidium, cesium, and ammonium phosphates are entirely water-soluble. Most other phosphates are merely soluble or are insoluble in water. As a regulation, the hydrogen and dihydrogen phosphates are somewhat more soluble than the analogous phosphates. The pyrophosphates are mostly water-soluble.

OCCURRENCE OF PHOSPHATES

Phosphates are the naturally occurring form of the constituent phosphorus, invent in numerous phosphate raw materials. In mineralogy and geology, phosphate deliberates to a rock or ore comprising phosphate ions. Inorganic phosphates are excavated to achieve phosphorus for consumption in agriculture and industry [1].

The major global manufacturer and exporter of phosphate is Morocco. Within North America, the chief deposits lie in the Bone Valley region of central Florida, the Soda Springs province of southeastern Idaho, and the coast of North Carolina. Slighter deposits are situated in Montana, Tennessee, Georgia, and South Carolina. The small landmass nation of Nauru and its neighbor Banaba Island, which cast-off to have gigantic phosphate credits of the finest feature, have been mined exceptionally. Rock phosphate can also originate in Egypt, Israel, Western Sahara, Navassa Island, Tunisia, Togo, and Jordan, countries that have enormous phosphate-mining industries.

Early in the 17th century, the word phosphate *phosphor* was invented and its meaning remains unchanged. It is said that an alchemist, Vincentinus Casciarolo of Bologna, Italy, found a heavy crystalline stone with a gloss at the foot of a volcano, and fired it in a charcoal oven intending to convert it to noble metal. This stone was called the "Bolognian stone" which appears to have been barite ($BaSO_4$) [2], with the fired product being BaS, which is at the present known to be a phosphate host for phosphor materials (Figure 1).

Figure 1. A milestone in the development of phosphors (Bolognian stone).

PHOSPHATES AND ITS CLASSIFICATION

Phosphates are compounds that contain oxyanions of phosphorus (V), ranging from the simple orthophosphate group to condensed chain, ring, and network anions. Oxyanions of phosphorus in lower oxidation states such as phosphite, HPO_3^{2-} are also known. A very large number of solid phosphates have been prepared or found as minerals. Their diversity results from variations in the phosphate species, the large number of cations to which they may be coordinated, and the presence of other anions or molecules, notably H_2O. Their chemistry is similar to that of solid silicates and borates. Much attention is focused on phosphates of metallic elements and other small cations (H^+, NH^{4+}), although a variety of phosphate salts of large organic or inorganic coordination complex cations are also known. Background information about the chemistry of phosphates and related phosphorus species exist in the texts by Corbridge [2] and Kanazawa [3]

Phosphate structures are generally rigid, resistant to chemical attack, and (when anhydrous) insoluble and thermally stable. This leads to some applications as nuclear waste immobilization hosts or negative thermal expansion materials. Many solid phosphate hosts permit diffusion of extra-framework species leading to potential uses as ion exchangers and conductors, and as microporous catalysts. Phosphate anions do not absorb significantly in the UV-visible region and so solid phosphates can also find use as optical materials such as glasses, phosphors, non-linear media, and lasers. Solid phosphates constitute many minerals, notably Apatites which are also found in living organisms as rigid components such as bones and teeth. Amorphous Phosphorite deposits are important sources of phosphate fertilizers. Depending upon the no of oxyanions they contain solid phosphates that are conveniently classified as given below.

Orthophosphates

The orthophosphate group, PO_4^{3-}, (often shortened to 'phosphate') is the most ubiquitous oxyanion of phosphorus. In solid orthophosphates, all four oxygen atoms are usually coordinated to cations resulting in a strongly bonded three-dimensional framework, although layered or chain structures sometimes result. The covalent bonding in the tetrahedral PO_4^{3-} anion may be described as the average of four resonance hybrids such as (1), giving the average structure (2) with tetrahedral (Td) symmetry. These two views of the bonding illustrate important features of phosphate chemistry. (1) shows that up to three covalent (P)O-X bonds may be formed with high valent elements X, notably P^V in condensed phosphates, whereas (2) shows that all four oxygen atoms are equally involved in predominantly ionic bonding in metal orthophosphates. In ionic phosphates, terminal P=O and P-O- bonds within the same tetrahedral group are equivalent and are strengthened by $P:3d\pi\text{-}O:2p\pi$ overlap. The orthophosphate group usually displays a near-regular tetrahedral geometry. Analysis of geometric data from 85 reliably determined crystal structures gives a mean P-O bond length of 1.536 Å with distances lying in the range 1.50-1.58 Å, and tetrahedral angles between 97 and 115°. Distorted geometries can occur when the orthophosphate group acts as a bidentate ligand, resulting in a strained four-membered ring. (3) shows the geometry of the ring formed by phosphate tetrahedra and CrO_6 octahedra sharing a common edge in a-$CrPO_4$, as determined by low-temperature neutron diffraction. The acid orthophosphate anions, (mono) hydrogenphosphate HPO_4^{2-} and dihydrogenphosphate H_2PO_4 also have extensive solid-state chemistries. Protonation lowers the bonding symmetry as the P-O(H) bonds have single bond character and the unprotonated oxygens are consequently bonded more strongly to phosphorus. Analysis of 21 acid orthophosphate structures gives P-O(H) distances in the range 1.56-1.62 Å,[4] whereas the average P-Ot bond distance (Ot = terminal oxygen) of 1.52 Å is slightly shorter than that in orthophosphates. The unprotonated oxygen atoms are readily coordinated to cations whereas the protonated sites are often uncoordinated, but are hydrogen-bonded to nearby P-OH groups or other suitable species. This tends to result in more open structures with lower dimensionalities for acid orthophosphates than for orthophosphates.

Orthophosphates can found in diverse forms. Structural variety results from a large number of cations that form stable orthophosphates and the incorporation of additional molecules, notably water, or anions. Virtually every metallic element forms an orthophosphate, sometimes in a variety of oxidation states, e.g., from V^{II} to V^V in $NaV^{II}V^{III}_2(PO_4)_3$, $V^{III}PO_4$, $V^{IV}O(H_2PO_4)_2$, and V^VOPO_4. The former compound is one of many mixed valent orthophosphates. A very large number of mixed cation orthophosphates are also known; a complex example is $Mg_{21}Ca_4Na_4(PO_4)_{18}$. Hydrated phosphates are common and variation of the water content may be possible; $VO(HPO_4).nH_2O$ structures have been characterized for n = 1/2, 1, 2 (two forms), 3, and 4. Orthophosphates containing additional anions include $Ca_2(PO_4)F$, $Fe_2(PO_4)O, Ca_{10}(PO_4)_6S, LiMn(PO_4)(OH), Zr_2(WO_4)(PO_4)_2, Ca_5(PO_4)_2(SiO_4), Pb_3Mn(PO_4)_2(SO_4)$, and $Na_3Ca(SiO_3)PO_4$ (containing an infinite catenasilicate chain). Solid solutions involving substitutions of the cations or the orthophosphate group (e.g., for orthoarsenate, vanadate, and silicate groups) further extend the range of possible phases.

Polyphosphates

Linking phosphate tetrahedra into chains through two vertices results in polyphosphate anions, $PnO_{3n+1}^{(n+2)-}$, also known as oligophosphates. Finite chains containing up to six tetrahedra have been found in the solid-state. They become less common with increasing n. A large number of anhydrous and hydrated triphosphates have been characterized, including structures containing the mono- and di- hydrogentriphosphate anions.

Only the terminal phosphate groups are protonated, as bridging -OP(O$_2$H)O$^-$ groups have low pKa's. The layered triphosphates MH$_2$P$_3$O$_{10}$·2H$_2$O (M =Al, Cr, Mn, Fe) are intercalation hosts. Tetraphosphates are less common than triphosphates and the best-defined examples are crystalline anhydrous materials. Acid tetraphosphate, (NH$_4$)$_4$H$_2$P$_4$O$_{13}$, has been reported. Pentaphosphate anions have been structurally characterized in Mg$_2$Na$_3$P$_5$O$_{16}$, CsM$_2$P$_5$O$_{16}$ (M = V, Fe) and the mixed phosphate Rb$_2$Ta$_2$H(PO$_4$)$_2$(P$_5$O$_{16}$). One hexaphosphate, Ca$_4$P$_6$O$_{19}$, has been reported [4], but the structure has not been determined. There is evidence for longer polyphosphate anions up to at least P$_8$O$_{25}{}^{10-}$ in solution, but no well-defined solid derivatives have yet been prepared. Several solid structures containing two polyphosphate anions have been reported [4]. All have complex stoichiometries involving at least two cationic species, examples are K$_2$Ni$_4$(PO$_4$)$_2$P$_2$O$_7$, CsTa$_2$(PO$_4$)$_2$P$_3$O$_{10}$, NH$_4$Cd$_6$(P$_2$O$_7$)$_2$P$_3$O$_{10}$, andCaNb$_2$O(P$_2$O$_7$)P$_4$O$_{13}$, KAl$_2$(H$_2$P$_3$O$_{10}$)P$_4$O$_{12}$ contains the dihydrogen triphosphate and cyclotetraphosphate anions.

Cyclophosphates

Previously known as cyclopolyphosphates, these rings may contain up to 12 tetrahedra, but those with three, four, and six units are most common. The cyclotri- and cyclotetra- phosphate rings adopt puckered geometries typical of saturated six and eight atom rings. The predominance of even membered cyclophosphates reflects their ability to pack efficiently in the solid-state, rather than any inherent stability over odd membered ones. This is often reflected by a high internal symmetry in the crystalline state; an analysis of thirty reliably determined cyclohexaphosphate structures shows that eighteen have inversion symmetry and a further seven have 3-fold (D3d) internal symmetry. Both hydrated and anhydrous cyclophosphates have been prepared, but no acid anions have been found in these solids due to the low basicity of two-connected phosphate groups [4]. A common structural feature, especially with large rings, is the formation of layers of cyclophosphate groups. This enables a large range of hydration numbers to be observed, as cations can be coordinated between two layers in anhydrous salts, or by cyclophosphate groups on one side and water molecules on the other in some hydrated compounds. Fully hydrated cations can also lie between the layers and further non-coordinated water molecules may occupy the intra-annular and interlamellar spaces. Examples of highly hydrated cyclohexaphosphates are $Nd_2P_6O_{18}.12H_2O$ and $Cu_3P_6O_{18}.14H_2O$.

Catenaphosphates

The infinite chain catenaphosphate anion (PO_3^-) represents the infinite limit of the poly- and cyclophosphate series. Catenaphosphates are formed at high temperatures and so all reported structures are anhydrous, e.g., $Al(PO_3)_3$, $UO_2H(PO_3)_3$, and $Cs_2Co(PO_3)_4$. P-Ot and P-Ob distances are ~1.48 and ~1.60 Å, respectively, similar to values for two-connected phosphate tetrahedra in other anions. The cations lie between parallel, infinite polyphosphate chains resulting in strongly-bonded three-dimensional structures. Thermal decomposition of hydrogen phosphates has resulted in the acid catenaphosphates $Na_2H(PO_3)_3$, $BiH(PO_3)_4$ and $UO_2H(PO_3)_3$. In $BiH(PO_3)_4$, one of the terminal oxygens on every fourth phosphate tetrahedron is protonated; this is the only non-bridging oxygen not to coordinate with Bi^{3+}. As cyclophosphates and the catenaphosphate anion (together termed metaphosphates) share the basic composition (PO_3^-) n, crystal structure analysis is often the only way to determine which species is present, although IR spectroscopy may be useful. Different isomers of the same metaphosphate composition are often found. Six crystal forms (A-F) of $M(PO_3)_3$ (M = Al, Cr, Mn, Fe, Ga) have been identified, three of which have been structurally characterized and found to contain cyclotetraphosphate (form A), cyclohexaphosphate (B), and catenaphosphate (C) anions. However, structures containing two metaphosphate anions are very rare, an example is $Pb_2Cs_3(P_4O_{12})(PO_3)_3$ in which both cyclotetraphosphate and catenaphosphate groups are present.

Ultraphosphates

Possible ultraphosphate anions containing x two-connected and y three-connected tetrahedra have stoichiometry $P_{x+y}O_{3x+5y/2}{}^{x-}$. All observed anions have y = 2, and so maybe written $P_nO_{3n-1}{}^{(n-2)-}$, where n = 4, 5, 6, and 8. Solid ultraphosphates are anhydrous, as the P-O-P bridges between three-connected phosphate tetrahedra are susceptible to hydrolysis. The potential of lanthanide ultraphosphates MP_5O_{14} as laser materials has driven the exploration of ultraphosphate chemistry. The list of characterized ultraphosphates is; MP_4O_{11} (M = Mg, Ca, Mn, Co, Ni, Cu, Zn), MP_5O_{14} (M = Lanthanide, Y, Bi), $NiHP_5O_{14}$, $M_2P_6O_{17}$ (M =Ca, Cd, Sr), $(UO_2)2P_6O_{17}$, $(TaO_2)4P_6O_{17}$ and $Na_3MP_8O_{23}$ (M = Al, V, Cr, Fe, Ga). The structures of these anions vary from discrete anions to infinite ribbons, sheets, and three-dimensional frameworks. The MP_4O_{11} structures contain infinite layers of fused eight and twelve membered rings, but the MP_5O_{11} structures vary with the cation radius and fall into three principal types. Types I (M = La-Ho, Bi) and III (M = Dy-Lu, Y) both contain infinite ribbons (5), while in type II materials (M =Tb-Lu, Y) a complex three-dimensional phosphate framework is formed. A unique polymorph of CeP_5O_{11} contains infinite sheet anions. Sheets of fused fourteen membered rings are found in $M_2P_6O_{17}$ compounds, whereas $UO_2P_6O_{17}$ contains an infinite three-dimensional anionic network. The $Na_3MP_8O_{23}$ structure contains the unique cage phosphate anion $P_8O_{23}{}^{6-}$ (6) which has a three-fold symmetry axis. This is the only molecular ultraphosphate anion known to have been characterized. Phosphorus pentoxide may be regarded as a neutral ultraphosphate containing only three-connected phosphate tetrahedra. Two extended forms of P_2O_5 are known, one containing infinite sheets, and the other a three-dimensional framework. The third form, molecular P_4O_{10} (7), consists of four tetrahedra each joined through three vertices and is the most highly condensed phosphate species.

Substituted Anions

Condensing phosphate tetrahedra with other XO_4^{n-} groups results in P-substituted anions such as the polyphosphate derivatives $Na_3PS_2O_{10}$, $Na_4P_2S_2O_{13}$, $Na_3PCr_3O_{13}.3H_2O$, and $Li_3PCr_4O_{16}$. The phosphate groups are found in the bridging positions of these chains. Infinitely extended tetrahedral anions formed by condensing phosphate and other tetrahedral groups. Electronegative non-metals can substitute for oxygen in phosphate anions. Many fluoro- and difluoro-orthophosphates are known, e.g., $CaPO_3F.2H_2O$, $Cu(PO_2F_2)_2$, and sodium salts of all the thio-orthophosphate anions $PO_{4-n}S_n^{3-}$ (n = 1-4) have been crystallized. Condensed thiophosphates have also been reported, e.g., $Cs_3P_3O_6S_3$ and $Na_4P_4O_8S_4.6H_2O$, and analogous cycloimidophosphate anions are found in $K_3P_3O_6(NH)_3$ and $Cs_4P_4O_8(NH)_4.6H_2O$. In the cyclothiophosphates, one terminal sulfur atom is present on each phosphate unit, however, the NH groups occupy the bridging positions in the cycloimidophosphate rings. Many monovalent and divalent cation salts of the $P_3O_6(NH)_3^{3-}$ anion have been prepared [5]. $AgPO_2(NH_2)_2$ is a rare example of a diamidophosphate, the nitrogen analog of a dihydrogenphosphate.

A literature survey reveals that to this moment, plenty of phosphors have been reported, such as aluminates, silicates, sulfides, fluorides, vanadates, and oxynitrides /nitrides, but phosphors for solid-state lighting with commendable optical properties are still exceptionally rare [5,6]. As the aim of this work is to carry out a detailed investigation on the synthesis, structural, morphological and photoluminescence properties of the synthesized pyrophosphate phosphor, this work is especially concentrated on the phosphate-based phosphors amongst the entire family.

PYROPHOSPHATES/DIPHOSPHATES

The diphosphate anion, $P_2O_7^{4-}$ (also known as pyrophosphate) is the simplest polyphosphate anion and is found in many solids. Variations in composition and structure are similar to those described above for the orthophosphates. The bonding and consequent geometry of the Ob-PO_3 (Ob =bridging oxygen) groups are very similar to that of the (H)O-PO_3 group. P-Ob distances are in the range 1.58-1.64 Å whereas the reported mean P-Ot distance for 17 diphosphates is 1.512 Å. P-O-P angles lie in the range 120-160°. Unusually large values are a result of a disorder of the bridging oxygen atom, notably in the 'linear,' centrosymmetric diphosphate groups (4) present in Thortveitite type transition metal diphosphate structures, $M_2P_2O_7$. The disorder is evidenced by anomalous P-Ob distances and Ob-P-Ot angles, and a large amplitude of thermal vibration for the bridging oxygen. Careful refinement of $Mn2P_2O_7$ using single-crystal X-ray and powder neutron diffraction data shows that the bridging oxygens lie 0.2 Å on either side of the inversion center, giving a P-O-P angle of 166°, in better agreement with well-ordered diphosphate groups. In general, apparently linear P-O-P linkages in condensed phosphates arise from such disorder and do not reflect a stable geometry. All three acid diphosphate anions have been found in the solid-state, e.g., $Mn(HP_2O_7)$, $Na_2(H_2P_2O_7).6H_2O$ and $Cs(H_3P_2O_7).H_2O$. However, trihydrogendiphosphate salts are rare due to the low pKa of the doubly protonated phosphate group (cf H_3PO_4 has pK1 = 2.1). The bridging oxygen is of very low basicity and does not coordinate even to highly charged cations. This is a universal feature in the chemistry of condensed phosphates. The consequent geometry of pyrophosphate is shown in Figure 2.

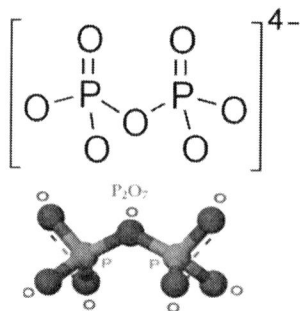

Figure 2. The bonding and consequent geometry of pyrophosphate.

Inorganic Pyrophosphatase

It is an enzyme that catalyzes the transformation of one ion of pyrophosphate to two phosphate ions [7]. This is a highly exergonic reaction, and consequently can be united to critical biological transformations to enterprise these transformations to end [8]. The functionality of this enzyme plays a precarious role in lipid metabolism calcium absorption and bone formation, [9, 10] and DNA combination, [11] as well as other biochemical alterations [12-13]. Two types of inorganic diphosphatase, identical in terms of both amino acid categorization and assembly, have been described to date: soluble and transmembrane proton-pumping pyrophosphatases (sPPases and H(+)-PPases, respectively). sPPases are ubiquitous proteins that hydrolyze pyrophosphate to generate heat, however, H+-PPases, so far unrevealed in animal and fungal cells, combine the energy of PPi hydrolysis to proton drive transversely biological membranes [14-16].

Applications of Pyrophosphate in Various Domains

In Biochemistry

The anion P_2O^{4-} is shortened PP$_i$, stand-up for inorganic pyrophosphate. It is molded by the hydrolysis of ATP into AMP in cells.

Introduction to Pyrophosphate

ATP → AMP + PP$_i$

For instance, when a nucleotide is integrated into a developing DNA or RNA component by a polymerase, pyrophosphate (PPi) is free. Pyrophosphorolysis is the reverse of the polymerization reaction in which pyrophosphate reacts with the 3'-nucleoside monophosphate (NMP or dNMP), which is disconnected from the oligonucleotide to relief the equivalent triphosphate (dNTP from DNA, or NTP from RNA). The pyrophosphate anion has the muster $P_2O^{4-}_7$, and is an acid anhydride of phosphate. It is unstable in aqueous solution and hydrolyzes into non-living phosphate:

$P_2O^{4-}_7 + H_2O \rightarrow 2HPO^{2-}_4$

or in ecologists' shorthand representation:

PP$_i$ + H$_2$O → 2 P$_i$ + 2 H$^+$

In the nonexistence of enzymic catalysis, hydrolysis responses of modest polyphosphates such as pyrophosphate, rectilinear triphosphate, ADP, and ATP usually progress very progressively in all but enormously acidic media [17].(The inverse of this reaction is a procedure of making pyrophosphates by the heating system of phosphates.)This hydrolysis to inorganic phosphate successfully reduces the cleavage of ATP to AMP and PP$_i$ irreversible, and biochemical reactions coupled to this hydrolysis are irreversible as well. PP$_i$ ensues in synovial fluid, blood plasma, and urine at levels adequate to block calcification and maybe a natural inhibitor of hydroxyapatite development in the extracellular fluid (ECF) [18]. Cells may channel intracellular PPi into ECF [19]. ANK is a nonenzymatic plasma-membrane PPi channel that supports extracellular PPi levels [19]. The substandard function of the membrane PPi channel ANK is connected with low extracellular PPi and elevated intracellular PPi [18]. Ectonucleotide pyrophosphatase/phosphodiesterase (ENPP) may purpose to raise extracellular PPi [6]. From the viewpoint of high energy phosphate accounting, the hydrolysis of ATP to AMP and PPi requires two high-energy phosphates, as to reconstitute AMP into ATP requires two phosphorylation reactions.

AMP + ATP → 2 ADP

2 ADP + 2 P_i → 2 ATP

The plasma concentration of inorganic pyrophosphate has a reference range of 0.58–3.78 µM (95% prediction interval) [20].

Terpenes

Isopentenyl pyrophosphate converts to geranyl pyrophosphate the precursor to tens of thousands of terpenes and terpenoids shown in Figure 3 [21].

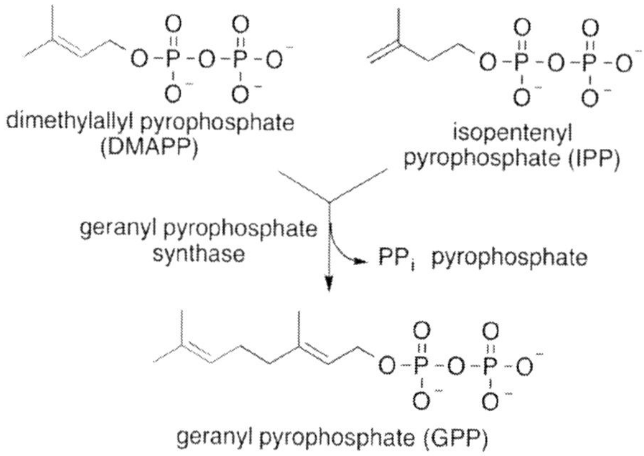

Source: https://en.wikipedia.org/wiki/Pyrophosphate#/media/File:Synthesis_of_geranyl_pyrophosphate.png.

Figure 3. Conversion of isopentenyl pyrophosphate to geranyl pyrophosphate.

Isopentenyl pyrophosphate (IPP) and dimethylallyl pyrophosphate (DMAPP) summarize to yield geranyl pyrophosphate, originator to all terpenes and terpenoids.

As a Food Additive

Various diphosphates are used as emulsifiers, stabilizers, acidity regulators, raising agents, sequestrants, and water retention agents in food processing [22]. They are classified in the E number scheme under E450 [23]:

- E450(a): disodium dihydrogen diphosphate; trisodium diphosphate; tetrasodium diphosphate (TSPP); tetrapotassium diphosphate
- E450(b): pentasodium and pentapotassium triphosphate
- E450(c): sodium and potassium polyphosphates

In particular, various formulations of diphosphates are used to stabilize whipped cream.

As an Efficient Phosphor in Solid-State Lighting

In recent years, the luminescent properties of phosphate based phosphors with varying oxyanions have been extensively investigated because of their excellent thermal and chemical stability [24-25] and development of luminescent materials doped with rare-earth (RE) ions for lighting is one of the interesting fields of research. Doat and coworkers [24] have been extensively examined the photoluminescent properties of europium activated pyrophosphates. Alkaline earth pyrophosphate doped with rare-earth (RE) ions has attracted research interests in the field of photoluminescence since they are suitable hosts with high chemical stability, which offers better homogeneity and lowers sintering temperature and also can produce plenty of crystal field environments imposed on emission centers. Few studies on RE-doped pyrophosphates have been reported in the literature. Pelova and Grigorov [25] reported the synthesis of europium-doped zirconium pyrophosphate (ZrP_2O_7:Eu) under the air atmosphere and at high temperature. The luminescent properties, especially the Eu^{2+}-doped diphosphate, such as $Ca_2P_2O_7$:Eu^{2+}, $Sr_2P_2O_7$:Eu^{2+}, $MgSrP_2O_7$:Eu^{2+} and $MgBaP_2O_7$:Eu^{2+} were found to be efficient phosphors in the violet-blue region [26]. Phosphate based phosphors consist of a range of compositional and structural possibilities (ultra, meta, pyro, and ortho) which are excellent in chemical and physical properties useful for specific technological and lighting applications, besides pyrophosphate hosts allows diffusion of additional-structure species approaching to potential uses as ion exchangers and conductors. A literature survey [27-30] reveals that pyrophosphate compounds activated with RE^{3+} or transition metal ions are considered as efficient candidates for luminescent materials. They have admirable luminescent, dielectric, semiconductor, fluorescent, magnetic and ion-exchange properties, consequently, their technological significance is increasing day by day [31].

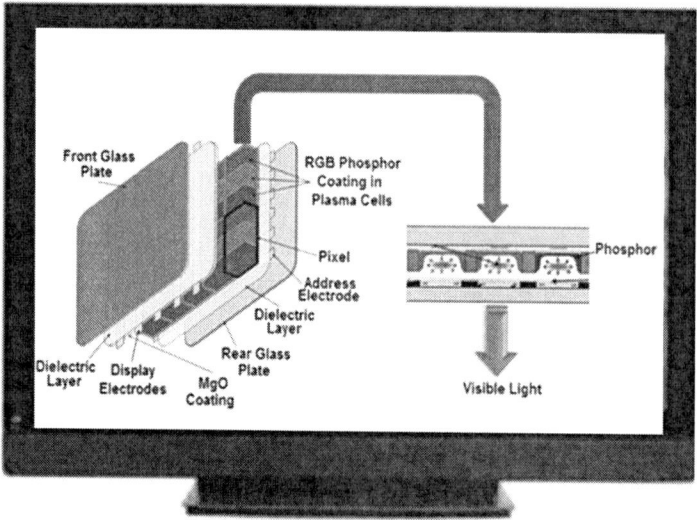

Figure 4. Schematic diagram of a plasma display and a plasma TV.

They are used in many modern portable devices such as laptops, cellular phones and digital cameras [32].

Some specific applications of pyrophosphate based hosts phosphors are:

- Plasma display panels (PDPs) (See Figure 4).
- Liquid crystal displays (LCDs)
- Organic light-emitting diode displays (OLEDs)
- Electroluminescent displays (ELDs)
- Surface-conduction electron-emitter displays (SEDs)
- Field emission displays (FEDs).

REFERENCES

[1] Hogan, C. Michael. 2011. *Phosphate. Encyclopedia of Earth. Topic ed. Andy Jorgensen.* Ed.-in-Chief C. J. Cleveland. National Council for Science and the Environment. Washington DC.

[2] Kanazawa, T. *Inorganic Phosphate Materials*, ed. Materials Science Monographs, 52, Elsevier, Amsterdam, 1989.

[3] Attfield, J. P. Phosphates: *Solid State Chemistry, Encyclopedia of inorganic chemistry-2nd Ed.* (2006) 1-52.

[4] Silver, J., R. Withnall, *Chem. Rev.* 104 (2004) 2833.

[5] Feldmann, C. *Adv. Funct. Mater.* 13 (2003) 511.

[6] Kuo, T. W., W. R. Liu, T. M. *Chen. Opt. Express.* 18 (2010) 1888.

[7] Harold, F. M. (December 1966). "Inorganic polyphosphates in biology: structure, metabolism, and function." *Bacteriological Reviews.* 30 (4): 772–94. PMC 441015. PMID 5342521.

[8] Terkeltaub, R. A. (July 2001). "Inorganic pyrophosphate generation and disposition in pathophysiology." American Journal of Physiology. *Cell Physiology.* 281 (1): C1–C11. doi: 10.1152/ajpcell.2001.281.1.C1. PMID 11401820.

[9] Orimo, H., Ohata, M., Fujita, T. (September 1971). "Role of inorganic pyrophosphatase in the mechanism of action of parathyroid hormone and calcitonin." *Endocrinology.* 89 (3): 852–8. doi:10.1210/endo-89-3-852. PMID 4327778.

[10] Poole, K. E., Reeve, J. (December 2005). "Parathyroid hormone - a bone anabolic and catabolic agent." *Current Opinion in Pharmacology.* 5(6):612–7. doi: 10.1016/j.coph.2005.07.004. PMID 16181808.

[11] Nelson, David L., Cox, Michael M. (2000). *Lehninger Principles of Biochemistry*, 3rd ed. New York: Worth Publishers. p. 937. ISBN 1-57259-153-6.

[12] Ko, K. M., Lee, W., Yu, J. R., Ahnn, J. (November 2007). "PYP-1, inorganic pyrophosphatase, is required for larval development and intestinal function in C. Elegans." *FEBS Letters.* 581(28): 5445–53. doi:10.1016/j.febslet.2007.10.047. PMID 17981157.

[13] Usui, Y., Uematsu T., Uchihashi T., Takahashi M., Takahashi M., Ishizuka M., et al. (May 2010). "Inorganic polyphosphate induces osteoblastic differentiation." *Journal of Dental Research.* 89 (5): 504–9. doi:10.1177/0022034510363096. PMID 20332330.

[14] Perez-Castineira, J. R., Lopez-Marques R. L., Villalba J. M., Losada M., Serrano A. (December 2002). "Functional complementation of yeast cytosolic pyrophosphatase by bacterial and plant H+-translocating pyrophosphatases." *Proc. Natl. Acad. Sci. U.S.A.* 99 (25): 15914–9. doi:10.1073/pnas.242625399. hdl:11441/26079. PMC 138539. PMID 12451180.

[15] Baltscheffsky, M., Schultz A., Baltscheffsky H. (September 1999). "H+ -PPases: a tightly membrane-bound family." *FEBS Lett.* 457 (3): 527–33. doi: 10.1016/S0014-5793(99)90617-8. PMID 10523139.

[16] Ho, A. M., Johnson M. D., Kingsley D. M. (Jul 2000). "Role of the mouse and gene in control of tissue calcification and arthritis." *Science.* 289 (5477): 265–70. doi: 10.1126/science.289.5477.265. PMID 10894769.

[17] Rutsch F., Vaingankar S., Johnson K., Goldfine I., Maddux B., Schauerte P., Kalhoff H., Sano K., Boisvert W. A., Superti-Furga A., Terkeltaub R. (Feb 2001). "PC-1 nucleoside triphosphate pyrophosphohydrolase deficiency in idiopathic infantile arterial

calcification." *Am J Pathol.* 158 (2): 543–54. doi: 10.1016/S0002-9440(10)63996-X. PMC 1850320. PMID 11159191.

[18] Ryan, L. M., Kozin F., McCarty D. J. (1979). "Quantification of human plasma inorganic pyrophosphate. I. Normal values in osteoarthritis and calcium pyrophosphate dihydrate crystal deposition disease." *Arthritis Rheum.* 22 (8): 886–91. doi: 10.1002/art.1780220812. PMID 223577.

[19] Eberhard Breitmaier (2006). "Hemi- and Monoterpenes." *Terpenes: Flavors, Fragrances, Pharmaca, Pheromones.* doi: 10.1002/9783527609949.ch2.

[20] *Codex Alimentarius 1A*, 2nd ed, 1995, pp. 71, 82, 91.

[21] Jukes, D. J., *Food Legislation of the UK: A Concise Guide*, Elsevier, 2013, p. 60–61.

[22] Molins, Ricardo A. *Phosphates in Food*, p. 115.

[23] Lan, Y., L. Yi, L. Zhou, Z. Tong, F. Gong, R. Wang, *Phys. B: Condens. Matter*, 405 (2010) 3489.

[24] Doat, A., F. Pelle, A. Lebugle, *J. Solid State Chem.* 178 (2005) 2354.

[25] Pelova, V. A., L. S. Grigorov, *J. Lumin.* 72–74 (1997) 241.

[26] Yen, W. M., M. J. Weber, *Inorganic Phosphors*, CRC Press, Boca Raton, 2004.

[27] Natarajan, V., M. K. Bhide, A. R. Dhoble, S. V. Godbole, T. K. Seshagiri, A. G. Page, Chung-Hsin Lu, *Mater. Res. Bull.* 39 (2004) 2065.

[28] Kohale, R. L., S. J. Dhoble. *AIP Conf. Proc.* 1391(2011) 203.

[29] Weil, M., R. Glaum, *Eur. J. Solid State Inorg. Chem.* L 35 (1998) 495.

[30] Khattak, G. D., E. E. Khawaja, L. E. Wenger, D. J. Thompson, M. A. Salid, A. B. Hallak, M. A. Daous, *J. Non-Cryst. Solids* 194 (1996) 1.

[31] Kohale, R. L., S. J. Dhoble, *Micro & Nano Letters*, 7(2012) 453.

[32] Hong, W. P., J. Singh, and P. Bhattacharya, *IEEE Trans.*, ED-7(1986)480.

In: A Closer Look at Pyrophosphates
Editors: Ritesh L. Kohale et al.
ISBN: 978-1-53617-730-5
© 2020 Nova Science Publishers, Inc.

Chapter 2

SYNTHESIS TECHNIQUES OF PYROPHOSPHATE PHOSPHOR MATERIALS

Dinesh Kumar[1] and Ram Sagar Yadav[2],*
[1]School of Materials Science and Technology,
Indian Institute of Technology, (Banaras Hindu University),
Varanasi, Uttar Pradesh, 221005, India
[2]Department of Zoology, Institute of Science,
Banaras Hindu University, Varanasi, Uttar Pradesh, 221005, India

ABSTRACT

The pyrophosphate phosphors are very interesting materials because they display important properties used in various applications, such as light emitting diodes, supercapacitors, catalytic materials, electrode materials for batteries, ionic solid electrolyte batteries, nonlinear-optical and dielectric materials, etc. Physical and chemical properties of a material strongly depend on synthesis route, particle size and crystallinity. Different

---
* Corresponding Author's E-mail: ramsagaryadav@gmail.com.

synthesis techniques provide different platforms for the synthesis of the various forms of the materials distinctly, such as single crystalline, polycrystalline, thin films, etc. This implies that synthesis methods are very interesting and important parameters for deciding the properties of a material. The present chapter briefly describes the basic properties and structures of the pyrophosphate phosphor materials. In this chapter, we have discussed the novelty of the pyrophosphate phosphor materials $[(A_x)^{4+}(P_2O_7)^{4-}]$, which can accommodate different cations at A-site. We have also discussed various routes such as solid, liquid and gas phase synthesis methods for the production of pyrophosphate phosphor materials mostly in the oxide powder form. We have briefly discussed the effect of synthesis techniques on the morphology and photoluminescence intensity of the pyrophosphate phosphor materials.

1. INTRODUCTION

In recent years, the pyrophosphate phosphor materials have attracted much concentration due to their potential applications in the various fields, for example light-emitting diodes (LEDs) [1-2], ionic solid electrolyte batteries [3-4], supercapacitors [5-6] and nonlinear-optical (NLO) behaviors [7]. Pyrophosphates are usually very important materials in biochemistry. The pyrophosphate phosphor materials are also important candidates for energy applications used as cathode materials in batteries [8-10]. In general, the pyrophosphate phosphor materials have two phosphorus atoms linked in the form of P-O-P, i.e., $(P_2O_7)^{4-}$ with common chemical formula $(A_x)^{4+}[P_2O_7]^{4-}$, where A is a metal cation and x can be taken as 4, 2 or 1 for mono-, di- or tetra-valent cations, respectively. The pyrophosphates are often known as diphosphates. The pyrophosphates are synthesized by heating of the phosphates; hence they are named as pyro-phosphates and they are generally white or colorless. An interactive chemical molecular structure of pyrophosphate is shown in Figure 1.

The uniqueness of the pyrophosphate materials is that they can accommodate verities of cations at A-site. The pure (undoped) pyrophosphate phosphor materials $(A_x)^{4+}[P_2O_7]^{4-}$ do not always provide desired properties. In order to make it useful, the doping at A-site is required. Doping at A-site improves various properties and also generates very interesting phenomena due to changes in the crystal structure, bond-lengths, ionic states, etc. The general chemical formula of A-site doped pyrophosphate phosphor materials can be written in the form of $(A_{x-y}A'_y)^{4+}[P_2O_7]^{4+}$ ($0 < y < 1$), where A' is the doped cation. Recently, lots of works have been carried out on rare-earth, alkali and alkaline earth doped pyrophosphate phosphor materials. Zhang et al. reported very interesting multicolor emission in Eu^{3+} and Tb^{3+} co-doped $MgIn_2P_4O_{14}$ pyrophosphate synthesized by solid state method. Host, singly and co-doped $MgIn_2P_4O_{14}$ pyrophosphates crystallize into monoclinic crystal structure with $C2/c$ space group. They observed that by adjusting doping concentrations of Eu^{3+} and Tb^{3+} ions in the $MgIn_2P_4O_{14}$ pyrophosphate, the color-tunable emission has been observed ranging from reddish to orange to yellow to green to blue regions [11].

Song et al. have studied structural, thermal and optical properties of newly synthesized single-crystalline and polycrystalline materials of $CsNaZnP_2O_7$, $RbNaZnP_2O_7$ and $RbLiMgP_2O_7$ pyrophosphate phosphors. They have found that $CsNaZnP_2O_7$ and $RbNaZnP_2O_7$ phosphors crystallize into monoclinic crystal structure with $P2_1/n$ space group while $RbLiZnP_2O_7$ crystallizes into orthorhombic structure with $Pnma$ space group. Thermal analysis of these phosphors revealed that $CsNaZnP_2O_7$, $RbNaZnP_2O_7$ and $RbLiMgP_2O_7$ pyrophosphate phosphors have melting point around 781, 738 and 757°C, respectively. Theoretical calculation showed indirect band gaps of 3.75, 3.69 and 4.40 eV for $CsNaZnP_2O_7$, $RbNaZnP_2O_7$ and $RbLiMgP_2O_7$ phosphors, respectively [12]. Mahajan et al. have synthesized a series of Dy^{3+} doped $Mg_2P_2O_7$ pyrophosphate phosphors using solution combustion technique and studied their structural and optical behaviors. Structural analyses of these compounds revealed that these compounds have monoclinic crystal with $B2_1/c$ space group. The optical band gap of the Dy^{3+} doped $Mg_2P_2O_7$ phosphors was found to be 5.4 eV. They have suggested that the Dy^{3+} doped $Mg_2P_2O_7$

phosphors can be used for white LEDs exciting at 348 nm (by n-UV source) and have prospective applications in solid state lighting (SSL) [13]. Recently, Marje et al. have prepared thin films of hydrous nickel pyrophosphate on stainless steel substrate and discussed its application in electrocatalytic activity [14].

The crystal structure of the pyrophosphate phosphors is not unique. They crystallize into various crystal structures and have different space groups for the same crystal structure. In general, the pyrophosphate phosphor materials crystallize into cubic, tetragonal, orthorhombic, monoclinic, triclinic, etc. crystal structures [8, 15]. For example, $Na_2VP_2O_7$ [8] and $Li_2Fe_{1-3x/2}V_xP_2O_7$ [10] crystallize into monoclinic structure with space group $P2_1/c$; α-$Na_2CuP_2O_7$ and β-$Na_2CuP_2O_7$ crystallize into monoclinic structure with $P2_1/n$ and $C2/c$ space groups at room temperature and at higher temperature, respectively [16]; while $CsNaZnP_2O_7$ and $RbNaZnP_2O_7$ crystallize into monoclinic structure with space group $P2_1/n$ [12]; $Na_{1.56}Fe_{1.22}P_2O_7$ [4], $Na_2MnP_2O_7$ and $Na_2FeP_2O_7$ [8] crystallize into triclinic structure with space group P-1 or P1 and $Na_2CoP_2O_7$ [8] crystallizes into tetragonal and orthorhombic structures with $P4_2/mmm$ and $Pna2_1$ space groups, respectively. $RbLiZnP_2O_7$ pyrophosphate crystallizes into orthorhombic structure with space group $Pnma$ [12]. At room temperature, ZrP_2O_7 displays an orthorhombic structure with $Pbca$ space group, whereas at high temperature it shows cubic structure with $Pa\bar{3}$ space group [15, 17]. Figure 2 shows some unit cell molecular structures for triclinic, monoclinic, orthorhombic, tetragonal and cubic crystal structures to view the interactive pictures.

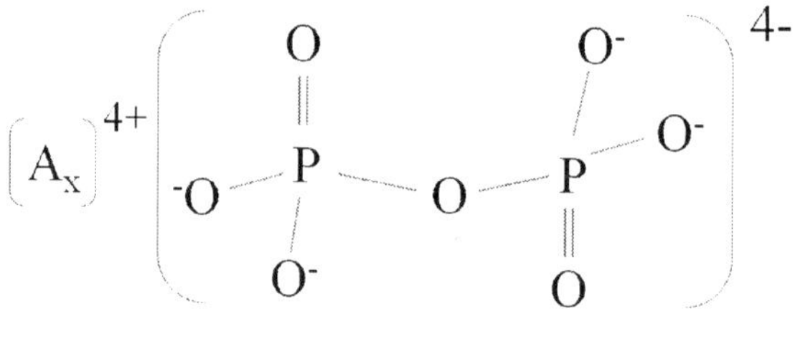

Figure 1. Interactive chemical molecular structure of pyrophosphate compound.

Synthesis Techniques of Pyrophosphate Phosphor Materials 25

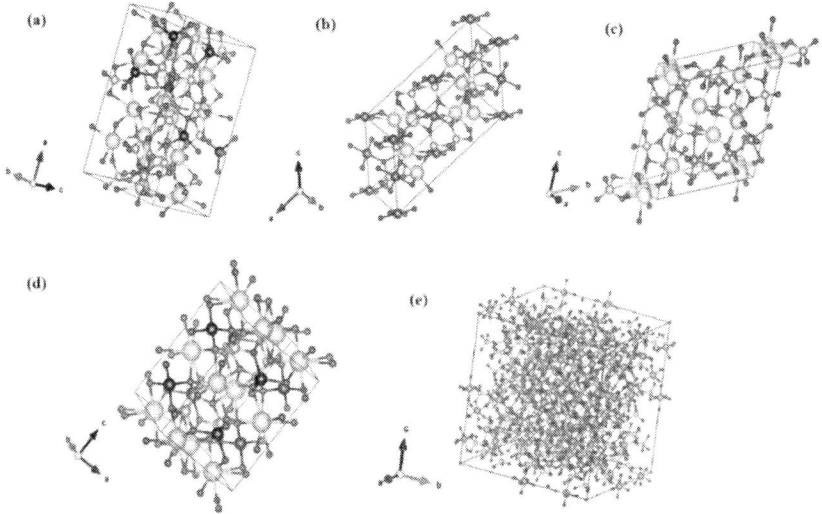

Figure 2. Ball and stick molecular model for (a) orthorhombic ($Na_2CoP_2O_7$), (b) monoclinic ($Na_2CuP_2O_7$), (c) triclinic ($Na_2FeP_2O_7$), (d) tetragonal ($Na_2CoP_2O_7$), and (e) cubic (ZrP_2O_7) crystal structures. The ball with blue color stands for Co (a and d) and Cu (b), yellow color stands for Na (a-d), white color stands for Fe (c), red color stands for O (a-e), purple color stands P (a-e) and green color stands for Zr (e) elements.

In this chapter, we have to discuss the synthesis processes used for the preparation of different types of alkali or alkaline earth or rare-earth doped pyrophosphate phosphor materials alongwith the impact of synthesis techniques on morphology and photoluminescence intensity of these materials.

2. SYNTHESIS TECHNIQUES FOR PYROPHOSPHATE PHOSPHOR MATERIALS

There are various techniques adopted for the synthesis of pyrophosphate phosphor materials. As we know that the properties of the pyrophosphate phosphor materials strongly depend on the synthesis routes. Therefore, the synthesis methods are very important factor to get the desired properties from pyrophosphate phosphor materials. These synthesis techniques can be divided into the following categories:

1. Solid phase synthesis
2. Liquid phase synthesis
3. Gas phase synthesis

2.1. Solid Phase Synthesis

The solid phase/state synthesis technique is a widely used method for the synthesis of polycrystalline solid pyrophosphate phosphor materials from the mixture of solid raw substances. The starting materials do not react to each other at room temperature. When it is fired at higher temperatures i.e., 800-1500°C the reaction occurs at a remarkable rate. The solid state synthesis technique requires raw materials in the oxide forms.

2.1.1. Mechanical Ball-Milling Method

It is a solid phase synthesis technique for the preparation of polycrystalline pyrophosphate phosphor materials widely used by researchers. This method involves grinding and mixing of raw materials in several times. For the synthesis of any pyrophosphate phosphor materials, one has to take stoichiometric amounts of metal oxides as starting materials. In this method, one can also use metal carbonates for the synthesis because at ordinary temperature, no chemical reaction takes place.

Let us discuss the entire process of this method by considering an example of $ALnP_2O_7$ (A = Cs, Rb, Tl and Ln = Lu, Tm, Y) pyrophosphate phosphors [18]. At first, the stoichiometric amounts of rare earth oxide [Lu_2O_3, Tm_2O_3 and Y_2O_3] alkali metal carbonate [Cs_2CO_3, Rb_2CO_3 and Tl_2CO_3] and ammonium dihydrogen phosphate [$NH_4H_2PO_4$] were weighed and grinded in mortor using pestle by hand for 1-2 hours. After the hand grinding of all the raw materials, the mixtures were ball milled in a planetary ball milling system to further mixing in the presence of acetone or alcohol as mixing medium for 6-8 hours at a normal rpm of 50-80 in clockwise and anticlockwise directions. After ball milling, the mixtures were dried and divided into several parts for calcination at different temperatures (i.e., 900-1500°C) for different time intervals ranging 5-24 hours to optimize the pure phase of $ALnP_2O_7$ pyrophosphate phosphors. Mbarek et al. have synthesized $ALnP_2O_7$ (A = Cs, Rb, Tl and Ln = Lu, Tm, Y) pyrophosphate phosphors using present method following mixing and ball-milling of raw materials and pre-calcination at 400°C for 4 hours in order to decompose $NH_4H_2PO_4$ and then calcined at 720°C for 20 hours. To measure properties, such as dielectric, piezoelectric and impedance measurements, the calcined powder of the prepared sample can be converted into a pellet form and sintered at a higher temperature than calcination temperature to get highly dense material [19]. Materials synthesized by this method have particles size in the micrometer (µm) range [11]. Some other rare earth doped $Mg(In_{1-x-y}Eu_xTb_y)_2P_4O_{14}$ ($MgIn_2P_4O_{14}$: Eu_x, Tb_y, $x = 0$-0.07, $y = 0.005$-0.11); Eu^{2+} and Ce^{3+} doped $SrCaP_2O_7$; and $AgYP_2O_7$ doped with Tb^{3+}, Sm^{3+} and Dy^{3+} (x = 0, 1, 3 5, 10, 15, 20 mol%), etc. pyrophosphate phosphors were also synthesized using mechanical ball-milling technique of solid-state synthesis method [11, 20-21]. In these cases, the particles size of these phosphor materials is also in µm range.

2.1.2. High Energy Ball-Milling Method

The preparation of pyrophosphate phosphor materials using this technique is very similar to that of Mechanical Ball-Milling technique. It is used for the synthesis of nanoparticles of the oxide materials. This method strictly bounds to use only metal oxides for the synthesis because there is a possibility of chemical reaction during high energy ball-milling, which may generate different toxic and/or unwanted gases. This technique adopts relatively large rpm from a few hundred to a few thousand. This method requires moderately low temperature for synthesis of the materials.

Now, we discuss the entire procedure of current synthesis method by taking an example of Eu^{3+}-doped $SrZnP_2O_7$ pyrophosphate phosphor. The stoichiometric amounts of highly pure SrO, ZnO, Eu_2O_3 and $NH_4H_2PO_4$ were taken as starting materials for the synthesis of Eu^{3+}-doped $SrZnP_2O_7$ pyrophosphate phosphor and were hand mixed using mortar and pestle. The starting materials were ball-milled for 5-24 hours with zirconia balls in presence of alcohol as mixing media in the 1:10 weight ratio of sample and balls of zirconia at 500-1000 rpm. The ball mixed material was dried at 90°C and was calcined at different temperatures (i.e., 500-900°C) and the XRD patterns were recorded to check the formation of pure phased sample.

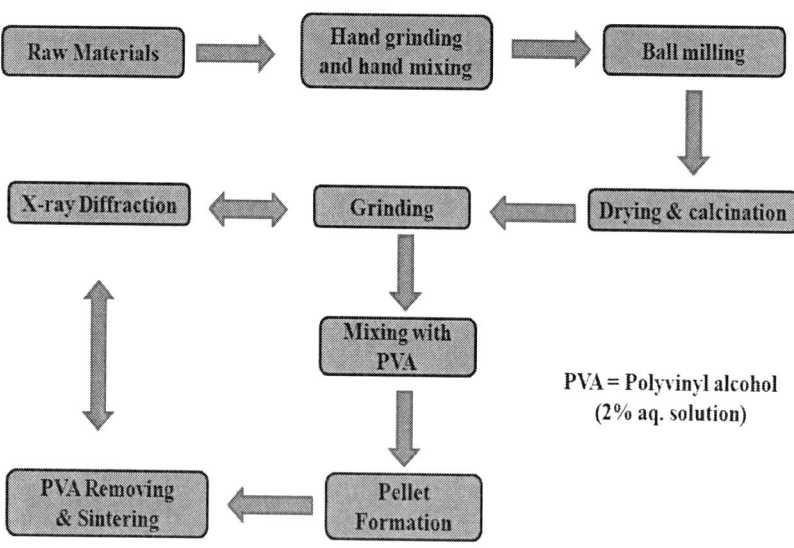

Figure 3. Flow-chart for the synthesis of oxide materials using solid state technique.

For further characterization, such as dielectric, piezoelectric and impedance measurements, etc. one can make pellets of the calcined powder sample by mixing in 2% aqueous solution of PVA (polyvinyl alcohol) used as binder. The formed pellet was fired at 500°C for 10 hours to remove the binder and sintered at higher temperature (say 1000°C) than calcination temperature to get the desired highly dense material. After sintering, we further recorded the XRD patterns to confirm the formed phase [22]. Figure 3 demonstrates flow chart of the synthesis procedures for the preparation of pyrophosphate phosphor materials using solid phase method. One can also use this method for the reduction of particles size from a micrometric scale to a nanometric scale.

2.2. Liquid Phase Synthesis

The liquid or chemical phase synthesis technique is widely used for the synthesis of nano-sized materials. In this method, the starting materials may react to each other at room temperature. There are various methods for the synthesis of pyrophosphate phosphor nano-materials in liquid phase synthesis, such as combustion, sol-gel, co-precipitation, hydrothermal methods, etc. and they are summarized below in more detail [15, 23, 31, 39].

2.2.1. Combustion Method
The combustion synthesis technique is a very facile and inexpensive method for the preparation of pyrophosphate phosphor nano-materials. This method requires raw materials in nitrate and/or acetate forms, which should be easily soluble in distilled water. It also requires an organic fuel such as glycine, urea or citric acid to help in the combustion process.

Let us consider the process of synthesis of the pyrophosphate phosphor materials using this method. Pazik et al. have synthesized Eu^{3+}-doped KYP_2O_7 pyrophosphate phosphor using highly pure raw materials, such as Eu_2O_3, Y_2O_3, $K_2CO_3.1.5H_2O$ and $(NH_4)_2HPO_4$; and urea (CH_4N_2O) used as fuel in stoichiometric amounts. Firstly, Eu_2O_3, Y_2O_3 and $K_2CO_3.1.5H_2O$ were dissolved in diluted nitric acid to convert into their respective nitrates and then $(NH_4)_2HPO_4$ and urea were dissolved into de-ionized water. All the precursor solutions of the raw materials were prepared separately and finally they were mixed together under continuous stirring and kept on a magnetic stirrer at 100-125°C and 200-225°C if glycine is used as used [23-24]. After constant stirring of 4-6 hours, the mixture solution turned thicker and forms gel, with further increase in time, an auto-ignition occurs resulting in flame, and generates huge amount of different gases. Finally, the mixed solution is converted into blackish-brown powder. At the time of ignition, the temperature of the whole reaction system may reach upto 800-1000°C for a very short duration [25]. The blackish-brown powder was collected and divided into several parts for the calcination at various temperatures from 550 to 900°C for 4 hours and the temperature is optimized to prepare a pure phased sample. Often, the variation in calcination/synthesis temperatures also results structural phase transition due to increase in the particles size. Pazik et al. have found β-KYP_2O_7 phase bearing monoclinic crystal structure with $P2_1/c$ space group below 700°C and above 700°C, they have obtained α-KYP_2O_7 phase having an orthorhombic crystal structure with *Cmcm* space group [23]. One can also prepare bulk samples of the same pyrophosphate phosphor materials by increasing calcination temperatures upto 1300°C [26]. Watras et al. have also synthesized RE^{3+} (RE = Sm, Tb and Dy) doped KYP_2O_7 pyrophosphate phosphors using this method [27]. It has been found that the phosphor materials prepared using this method produces nano-sized particles. Figure 4(a) displays a flow chart of the procedure adopted for the synthesis of pyrophosphate phosphor materials using this method.

2.2.2. Sol-gel Method

The sol-gel method involves physical and chemical procedures correlated with hydrolysis, condensation, polymerization, gelation, drying and densification [28]. This method also requires starting materials in the form of metal alkoxides. Metal alkoxides have the general chemical formula in the form of M(OR)x, which can be supposed as either a derivative of an alcohol ROH, where R is an alkyl group, in which the hydroxyl proton is substituted by a metal M or a derivative of a metal hydroxide M(OH)x [29]. The metal alkoxides of the desired metals are taken in the stoichiometric amounts and were dissolved in water or in a suitable solvent (say, alcohol) at an ambient temperature of 50-75°C under constant stirring. Controlling of pH value of the metal alkoxides solutions is imperative to keep away from the precipitation and to form the homogeneous gel that can be accomplished by adding acidic or basic solutions. These processes are known as hydrolysis and condensation leading to construction of the polymeric chains.

The formation of the polymeric chains ultimately leads to a noticeable improvement in the viscosity of the reaction mixture and the creation of a gel. The obtained gel was dried between 150-200°C to remove the volatile organic components and excess water, etc. from the gel [29-30]. After this, to obtain a single phase of the desired sample, the dried gel was calcined at various temperatures in the range of 500-800°C.

Kunghatkar et al. have synthesized a series of $CaMgP_2O_7:Gd^{3+}$ pyrophosphate phosphors using sol-gel method following several intermediate steps [31]. They used stoichiometric amounts of magnesium nitrate $Mg(NO_3)_2$, calcium nitrate $Ca(NO_3)_2$, gadolinium nitrate $Gd(NO_3)_3$, ammonium dihydrogen phosphate $NH_4H_2PO_4$, alongwith polyethylene glycol and citric acid ($C_6H_8O_7$) for the synthesis of series of $CaMgP_2O_7:Gd^{3+}$ pyrophosphate phosphors. All raw materials were dissolved individually in de-ionized water in separate beakers. All the solutions were mixed together on the hot plate under continuous stirring for 30 minutes. After this, a suitable amount of polyethylene glycol was added to the precursor solution used as polymerizing agent. The final solution was heated at 100°C for few hours to complete the polymerization and to evaporate the excess amount of solvents from the solution. Finally, the entire solution was converted into a viscous gel, which was calcined/annealed after drying at a temperature of 800°C for 12 hours [31]. Figure 4(b) shows a schematic flow chart of the processes involved in the synthesis of pyrophosphate phosphor materials using sol-gel method.

2.2.3. Co-precipitation Method

The co-precipitation method requires starting materials of metal cations from a common medium and precipitates in the form of carbonates, hydroxides, oxalates or citrates [32-35]. The obtained precipitates are consequently calcined at different temperatures to get a single phase of the desired product in powder form. Using this method, one can get a highly homogeneous powder sample. To achieve highly homogeneous material, the solubility of the products obtained during precipitation of metal cations should be very close to each-other [36]. In the co-precipitation process, the mixing of precursor solutions occurs at atomic level, which causes lower particles size of the final product and at very low calcination temperature to get the final product [37]. On the other hand, the synthesis of different materials requires particular conditions and precursor reactions, etc. It is to note that to obtain the ultimate product having desired properties, the co-precipitation method requires control of the pH of the solution, concentration, stirring speed and temperature of the mixture products [38].

This synthesis method can be understood by considering an example of $Mg_2P_2O_7:Eu^{3+}$ pyrophosphate phosphors as have been synthesized by Gupta et al. [39]. Figure 5 shows a schematic flow chart for the synthesis of $Mg_2P_2O_7:Eu^{3+}$ pyrophosphate phosphor using co-precipitation method. The stoichiometric amounts of magnesium nitrate $Mg(NO_3)_2$, diammonium hydrogen phosphate $(NH_4)_2HPO_4$, europium oxide (Eu_2O_3)

were used for the synthesis of $Mg_2P_2O_7:Eu^{3+}$ pyrophosphate phosphors. $Mg(NO_3)_2$ and $(NH_4)_2HPO_4$ were dissolved in the de-ionized water and Eu_2O_3 was dissolved in concentrated HNO_3.

All the precursor solutions were mixed together one by one under continuous stirring of the solutions at 50°C. The precursor solution of $(NH_4)_2HPO_4$ prepared in ethanol was added drop wise to the entire solution to precipitate a colorless powder of $Mg_2P_2O_7:Eu^{3+}$ pyrophosphate phosphor. The obtained precipitate was centrifuged at 2300 rpm for 30 minutes and dried at 80°C for 4 hours. Finally, the nano-particles of $Mg_2P_2O_7:Eu^{3+}$ pyrophosphate phosphor was obtained in the pure phase followed by annealing at 400°C for 4 hours in an open atmosphere.

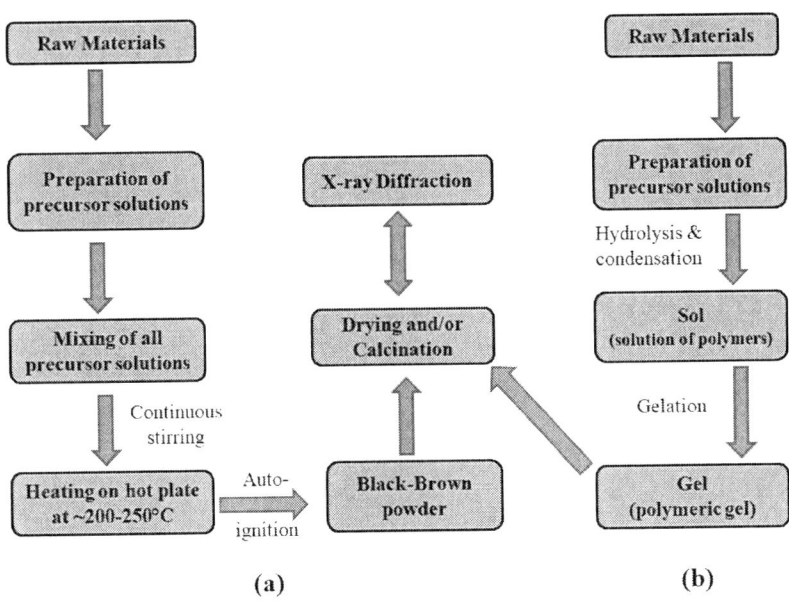

Figure 4. Flow chart for the synthesis of oxide materials using (a) combustion and (b) sol-gel methods.

2.2.4. Hydrothermal Reaction Method

Hydrothermal reaction method is also used to prepare the nano-materials. Let us introduce the entire process of synthesis of the nano-particles by considering an example of ZrP_2O_7 pyrophosphate phosphor synthesized by Samed et al. [15]. For the synthesis of ZrP_2O_7 pyrophosphate phosphor, they have followed three steps; (a) preparation of precursor of zirconium hydroxide; (b) hydrothermal reaction with phosphoric acid and (c) calcination in open atmosphere. To prepare the precursor of zirconium hydroxide, aqueous ammonia (NH_3)solution (10 M) was added slowly to 0.2 mol solution of zirconyl nitrate di-hydrate $ZrO(NO_3)_2 \cdot 2H_2O$ in 300 mL distilled water under continuous stirring until the pH of the solution reached to 10. A white gel is obtained, which was filtered and washed with de-ionized water many times and dried at 110°C for 12 hours. After this, obtained xerogel was dispersed in distilled water followed by stirring for 30 minutes and then H_3PO_4 acid was added drop wise slowly to this slurry under constant stirring for 6 hours. The entire mixture was moved to an inner-lined stainless steel autoclave having a 50 mL Teflon vessel and heated without stirring at 200°C for 18 hours. It was cooled to room temperature and a white gel is obtained, which was centrifuged, washed with de-ionized water and dried overnight under vacuum. After drying the product, it was calcined at different temperatures to optimize the synthesis temperature for 5 hours to get a pure phase. Figure 6 shows a schematic flow chart for the synthesis of ZrP_2O_7 pyrophosphate phosphor using hydrothermal method.

As we discussed that liquid phase synthesis techniques are used to synthesize nano-crystalline pyrophosphate phosphor materials at very low temperature. By using higher firing temperature, one can also produce bulk pyrophosphate phosphor materials in the μm sized particles, like solid phase synthesis route.

Synthesis Techniques of Pyrophosphate Phosphor Materials 35

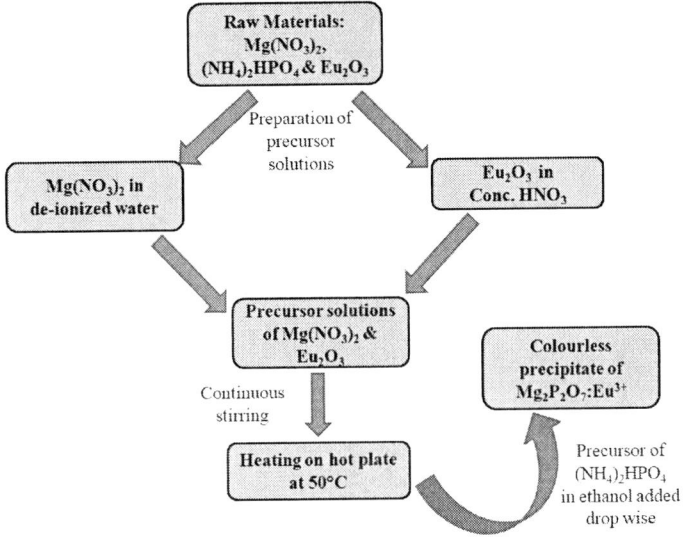

Figure 5. Flow charts for the synthesis of $Mg_2P_2O_7:Eu^{3+}$ pyrophosphate phosphor using co-precipitation method.

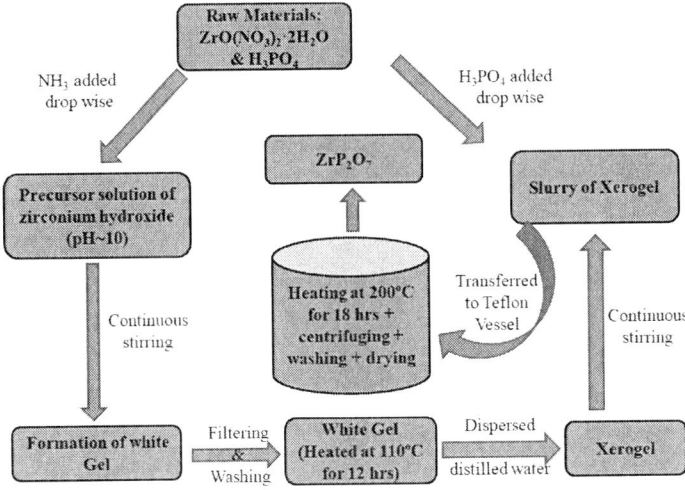

Figure 6. Flow charts for the synthesis of ZrP_2O_7 pyrophosphate phosphor using hydrothermal method.

2.3. Gas Phase Synthesis

Gas phase synthesis technique contains different methods, such as flames, furnaces, plasmas and lasers for the synthesis of powder oxide materials. In each method, the kinetics and thermodynamics of the reaction are similar but the designed reactor is different. To get the narrow particle size distribution of the oxide materials using gas phase synthesis, the dispersion must be minimized as it leads to an enlargement of the particles size [40]. Using gas phase reactors, one can produce highly pure pyrophosphate phosphor materials since it is comparatively simple to get purified reactant gases with impurities from the ppm (parts per million) to ppb (parts per billion) level.

Various electronic devices of phosphate phosphor materials are fabricated in the form of thin films using gas phase synthesis techniques, which is different from the synthesis methods used for the preparation of powder form of phosphate phosphors. Several techniques were developed for the preparation of thin films, such as chemical vapor deposition [41], plasma enhanced atomic layer deposition [42], magnetron sputtering [43], etc. The gas phase synthesis requires special arrangements and instrumentations for the preparation of good quality samples with the desired properties. The gas phase method can be classified into three categories:

1. Synthesis at the crystallization temperature under an appropriate atmospheric condition's temperature.
2. Synthesis in an intermediate temperature in the range of 600-800°C and then post-annealing treatment at higher temperatures.
3. Synthesis at very low substrate temperature and then post-annealing at very high temperature.

3. EFFECT OF SYNTHESIS TECHNIQUES ON MORPHOLOGY AND PHOTOLUMINESCENCE INTENSITY

The synthesis techniques of the pyrophosphate phosphor materials strongly affect the morphology of the materials and the photoluminescence (PL) emission intensity. Synthesis techniques also affect the crystallinity and crystal structure of the synthesized pyrophosphate phosphor materials. Kunghatkar et al. have studied the comparative properties of Eu^{3+} doped $CaMgP_2O_7$ pyrophosphate phosphors synthesized by four different methods using various characterization tools [31]. They have prepared $CaMgP_2O_7$ pyrophosphate phosphor doped with 1, 2, 5 and 7 mol% of Eu^{3+} ion using solid-state reaction, sol-gel, wet chemical and solution combustion methods. They prepared Eu^{3+} doped $CaMgP_2O_7$ phosphor materials by annealing them at 800°C temperature for all samples. Morphological analysis of the prepared materials using scanning electron microscopic (SEM) micrographs shows that the average particle size is bigger for the materials prepared using solid-state reaction. The particles of the Eu^{3+} doped $CaMgP_2O_7$ phosphor materials are in the irregular shape of rod for the samples prepared by solid-state reaction method, fine spherical particles for sol-gel method and agglomerated particles for wet chemical method and solution combustion method, respectively. The study of the emission spectrum of Eu^{3+} doped $CaMgP_2O_7$ phosphors monitoring at λ_{em} = 314 nm shows that the samples of Eu^{3+} doped $CaMgP_2O_7$ phosphors prepared by sol-gel method display largest PL intensity while it decreases for sol-gel, solution combustion, wet chemical and solid-state reaction methods sequentially. Table 1 shows a comparative data of average particles size, their shape and PL intensity [31].

Table 1. A comparison of average particles size, their shape and PL intensity of the Eu^{3+} doped $CaMgP_2O_7$ pyrophosphate phosphors

Synthesis Method	Range of average particle size	Shape of the particles	PL Intensity (arb. unit)
Solid-state reaction	5 µm	Irregular rod	42
Sol-gel method	2 µm	Fine spherical	370
Wet chemical method	2 µm	Agglomerated diffused shape	118
Solution combustion method	2 µm	Agglomerated diffused shape	270

On the other hand, the Ce^{3+}, Dy^{3+} co-doped $MgIn_2P_4O_{14}$-based phosphors prepared by solid state reaction method give intense fluorescence due to energy transfer from Ce^{3+} to Dy^{3+} under 337 nm excitation and have potential application in LEDs and fluorescent lamps [2]. The Eu^{3+}/Tb^{3+} doped and co-doped $Mg(In_{1-x-y}Eu_xTb_y)_2P_4O_{14}$ pyrophosphate phosphors were synthesized by Zhang et al. using solid state reaction and found the diffused spherically shaped particles in the range of few tens of micrometers [11]. They have also observed color-tunable emission with the concentrations of Eu^{3+} and Tb^{3+} ions at In-site in the $MgIn_2P_4O_{14}$ phosphors and the color tuned from reddish orange-yellow-green and blue colors. This phosphor found its application in fluorescent lamps. The $Mg_2P_2O_7$:Eu^{3+} nanocrystalline pyrophosphate phosphor was synthesized by Gupta et al. using co-precipitation method [39]. The $Mg_2P_2O_7$:Eu^{3+} pyrophosphate phosphor displays maximum PL intensity at 595 nm for 1 mol% Eu^{3+} doped $Mg_2P_2O_7$ upon 260 nm excitation wavelength and it can be used in solid state lighting purposes. Kohale et al. [44] have prepared Ce^{3+} and Dy^{3+} co-doped $SrZnP_2O_7$ pyrophosphate phosphor using solution combustion method. They have found agglomerated particles of irregular shaped with particles size in the range of few hundred nanometers. They have observed most intense emission peak at 388 nm for 5 mol% Ce^{3+} doped $SrZnP_2O_7$:Dy^{3+} phosphor on excitation with 327 nm and discussed its application in scintillating purposes. Some other phosphor materials were also synthesized by combustion method and solid state reaction method and they show their potential applications in displays devices, solid state lighting, color tunable based devices, bistable devices, solar cells, temperature sensing, etc. [45-50]. Thus, the synthesis method has significantly influenced the morphology and photoluminescence intensity of different pyrophosphate phosphor materials. One has to choose a particular synthesis method for obtaining the desired morphology, which can yield large photoluminescence intensity in the pyrophosphate phosphor materials.

4. CONCLUSION

This chapter summarizes the basic properties and structures of the pyrophosphate phosphor materials. We have discussed the novelty of the pyrophosphate phosphor materials, which accommodate different cations at A-site. We have also discussed various synthesis routes, such as solid, liquid and gas phase synthesis for the production of high quality pyrophosphate phosphor materials mostly in the oxide powder form. We have briefly discussed the effect of synthesis techniques on the morphology and the photoluminescence intensity of the pyrophosphate phosphor materials.

REFERENCES

[1] Fhoula, M; Dammak, M. Optical spectroscopy of thermal stable $Na_2ZnP_2O_7:Sm^{3+}/(Li^+, K^+)$ phosphors; *J. Lumin.*, 210, (2019), 1-6.

[2] Zhang, J; Zhang, GX; Cai, GM; Jin, ZP. Reduction of Ce(IV) to Ce(III) induced by structural characteristics and performance

characterization of pyrophosphate MgIn$_2$P$_4$O$_{14}$-based phosphors; *J. Lumin.*, 203, (2018), 590-598.

[3] Ng, MF; Sullivan, MB. First-principles characterization of lithium cobalt pyrophosphate as a cathode material for solid-state Li-ion batteries; *J. Phys. Chem. C*, 123, (2019), 29623-29629.

[4] Kosova, NV; Rezepova, DO; Petrov, SA; Slobodyuk, AB. Electrochemical and chemical Na$^+$/Li$^+$ ion exchange in Na-based cathode materials: Na$_{1.56}$Fe$_{1.22}$P$_2$O$_7$ and Na$_3$V$_2$(PO$_4$)$_2$F$_3$; *J. Electrochem. Soc.*, 164(1), (2017), A6192-A6200.

[5] Marje, SJ; Katkar, PK; Pujari, SS; Khalate, SA; Lokhande, AC; Patil, UM. Regulated micro-leaf like nickel pyrophosphate as a cathode electrode for asymmetric supercapacitor; *Synth. Met.*, 259, (2020), 116224-11.

[6] Chodankar, NR; Dubal, DP; Ji, SH; Kim, DH. Self-assembled nickel pyrophosphate-decorated amorphous bimetal hydroxides 2D-on-2D nanostructure for high-energy solid-state asymmetric supercapacitor; *Small*, 15, (2019), 1901145-11.

[7] Chen, J; Xiong, L; Chen, L; Wu, LM. Ba$_2$NaClP$_2$O$_7$: Unprecedented phase matchability induced by symmetry breaking and its unique fresnoite-type structure; *J. Am. Chem. Soc.*, 140, (2018), 14082-14086.

[8] Niu, Y; Zhangdand, Y; Xu, M. A review on pyrophosphate framework cathode materials for sodium-ion batteries; *J. Mater. Chem. A*, 7, (2019), 15006-15025.

[9] Shakoor, RA; Kahraman, R; Raja, AA. Thermal *insitu* analyses of multicomponent pyrophosphate cathodes materials; *Int. J. Electrochem. Sci.*, 10, (2015), 8941-8950.

[10] Xu, J; Chou, SL; Gub, QF; Din, MFM; Liu, HK; Dou, SX. Study on vanadium substitution to iron in Li$_2$FeP$_2$O$_7$ as cathode material for lithium-ion batteries; *Electrochim. Acta*, 141, (2014), 195-202.

[11] Zhang, J; Ma, ZY; Guo, JG; Cai, GM; Ma, L; Wang, XJ. Red-green-blue-tunable emission from Eu^{3+} and Tb^{3+} codoped pyrophosphate phosphors; *J. Lumin.*, 215, (2019), 116732, pp. 1-7.

[12] Song, H; Zhang, S; Li, Y; Liu, W; Lin, Z; Yao, J; Zhang, G. Syntheses, crystal structures, and characterizations of three new pyrophosphates $CsNaZnP_2O_7$, $RbNaZnP_2O_7$ and $RbLiMgP_2O_7$; *Solid State Sci.*, 95, (2019), 105940, pp. 1-7.

[13] Mahajan, R; Kumar, S; Prakash, R; Kumar, V; Choudhary, RJ; Phase, DM. X-ray photoemission and spectral investigations of Dy^{3+} activated magnesium pyrophosphate phosphors; *J. Alloys Compd.*, 777, (2019), 562-571.

[14] Marje, SJ; Katkar, PK; Kale, SB; Lokhande, AC; Lokhande, CD; Patil, UM. Effect of phosphate variation on morphology and electrocatalytic activity (OER) of hydrous nickel pyrophosphate thin films; *J. Alloys Compd.*, 779, (2019), 49-58.

[15] Samed, AJ; Zhang, D; Hinokuma, S; Machida, M. Synthesis of ZrP_2O_7 by hydrothermal reaction and post-calcination; *J. Ceram. Soc. Jpn.*, 119, (2011), 81-84.

[16] Erragh, F; Boukhari, A; Abraham, F; Elouadi, B. The Crystal Structure of α- and β-$Na_2CuP_2O_7$; *J. Solid State Chem.*, 12, (1995), 23-31.

[17] Birkedal, H; Andersen, AMK; Arakcheeva, A; Chapuis, G; Norby, P. |and Philip Pattison; The Room-temperature superstructure of ZrP_2O_7 is orthorhombic: There are no unusual 180° P-O-P bond angles; *Inorg. Chem.*, 4, (2006), 4346-4351.

[18] Mbarek, A. Synthesis, structural and optical properties of Eu^{3+}-doped $ALnP_2O_7$ (A = Cs, Rb, Tl; Ln = Y, Lu, Tm) pyrophosphates phosphors for solid state lighting; *J. Mol. Struct.*, 1138, (2017), 149-154.

[19] Krichen, M; Megdiche, M; Gargouri, M; Guidara, K. Frequency and temperature dependence of the dielectric properties and AC electrical conductivity in mixed pyrophosphate ceramic; *Ind. J. Phys.*, 88, (2014), 1051-1058.

[20] Shinde, KN. Luminescence in Eu^{2+} and Ce^{3+} doped $SrCaP_2O_7$ phosphors; *Results in Physics*, 7, (2017), 178-182.

[21] Hami, W; Daoud, M; Zambon, D; Elaatmani, M; Zegzouti, A. Synthesis and characterization of $AgYP_2O_7$ pyrophosphate activated with Tb^{3+}, Sm^{3+} and Dy^{3+} ions; *Inorg. Chem. Comm.*, 102, (2019), 192-198.

[22] Qin, L; Xu, C; Huang, Y; Kim, SI; Seo, HJ. Spectroscopic and structural features of Eu^{3+}-doped zinc pyrophosphate ceramic; *Cera. Inter.*, 40, (2014), 1605-1611.

[23] Pazik, R; Watras, A; Macalik, L; Deren, PJ. One step urea assisted synthesis of polycrystalline Eu^{3+} doped KYP_2O_7-luminescence and emission thermal quenching properties; *New J. Chem.*, 38, (2014), 1129-1137.

[24] Kumar, D; Singh, AK. Investigation of structural and magnetic properties of $Nd_{0.7}Ba_{0.3}Mn_{1-x}Ti_xO_3$ ($x = 0.05$, 0.15 and 0.25) manganites synthesized through a single-step process; *J. Magn. Magn. Mater.*, 469, (2019), 264-273.

[25] Patil, KC; Aruna, ST; Ekabaram, S. Combustion synthesis; *Curr. Opin. Solid State Mater. Sci.*, 2, (1997), 158-165.

[26] Kumar, D; Verma, NK; Singh, CB; Singh, AK. Crystallite size strain analysis of nanocrystalline $La_{0.7}Sr_{0.3}MnO_3$ perovskite by Williamson-Hall plot method; *AIP Conf. Proc.*, 1942, (2018), 050024, pp. 1-4.

[27] Watras, A; Deren, PJ; Pazik, R. Luminescence properties and determination of optimal RE^{3+} (Sm^{3+}, Tb^{3+} and Dy^{3+}) doping levels in the KYP_2O_7 host lattice obtained by combustion synthesis; *New J. Chem.*, 38, (2014), 5058-5068.

[28] Brinker, CJ; Scherer, GW. *Sol-gel science: The Physics and the chemistry of sol-gel processing*; Academic Press, Inc. London, (1990).

[29] Rahaman, MN. *Ceramic Processing and Sintering*; 2nd edition, Marcel Dekker, Inc. New York, (2003).

[30] Rajaeiyan, A; Mohagheghi, MMB. Comparison of sol-gel and co-precipitation methods on the structural properties and phase

transformation of γ and α-Al_2O_3 nanoparticles; *Adv. Manuf.*, 1, (2013), 176-182.

[31] Kunghatkar, RG; Barai, VL; Dhoble, SJ. Synthesis route dependent characterizations of $CaMgP_2O_7:Gd^{3+}$ phosphor; *Res. Phys.*, 13, (2019), 102295, pp. 1-9.

[32] Pei, RR; Chen, X; Suo, Y; Xiao, T; Ge, QQ; Yao, HC; Wang, JS; Li, ZJ. Synthesis of $La_{0.85}Sr_{0.15}Ga_{0.8}Mg_{0.2}O_{3-\delta}$ powder by carbonate co-precipitation combining with azeotropic-distillation process; *Solid State Ionics*, 219, (2012), 34-40.

[33] Uskokovic, V; Drofenik, M. Four novel co-precipitation procedures for the synthesis of lanthanum-strontium manganites; *Mater. Des.*, 28, (2007), 667-672.

[34] Cho, TH; Shiosaki, Y; Noguchi, H. Preparation and characterization of layered $LiMn_{1/3}Ni_{1/3}Co_{1/3}O_2$ as a cathode material by an oxalate co-precipitation method; *J. Power Sources*, 159, (2006), 1322-1327.

[35] Wei, Y; Han, B; Hu, X; Lin, Y; Wang, X; Deng, X. Synthesis of Fe_3O_4 nanoparticles and their magnetic properties; *Procedia Eng.*, 27, (2012), 632-637.

[36] West, AR. *Solid State Chemistry and its applications*; John Wiely Sons, USA, (2005).

[37] Gaikwad, AB; Navale, SC; Samuel, V; Murugan, AV; Ravi, V. A co-precipitation technique to prepare $BiNbO_4$, $MgTiO_3$ and $Mg_4Ta_2O_9$ powders; *Mater. Res. Bull.*, 41, (2006), 347-353.

[38] Zawrah, MF; Hamaad, H; Meky, S. Synthesis and characterization of nano $MgAl_2O_4$ spinel by the co-precipitated method; *Ceram. Int.*, 33, (2007), 969-978.

[39] Gupta, KK; Dhoble, NS; Burghate, DK; Dhoble, SJ. Luminescence properties of nanocrystalline $Mg_2P_2O_7:Eu$ phosphor; *Luminescence*, 33(5), (2018), 947-953.

[40] Ring, TA. *Fundamentals of ceramic powder processing and synthesis: Ceramic powder synthesis*, Academic Press, (1996).

[41] Goto, T. & Katsui, H. (2015). Chemical Vapor Deposition of Ca-P-O Film Coating; In: K. Sasaki, O. Suzuki, N. Takahashi (eds) *Interface Oral Health Science*, 2014, Springer, Tokyo.

[42] Dobbelaere, T; Mattelaer, F; Roy, AK; Vereecken, P; Detavernier, C. Plasma-enhanced atomic layer deposition of titanium phosphate as an electrode for lithium-ion batteries; *J. Mater. Chem. A*, **5**, (2017), 330-338.

[43] Yonggang, Y; Wolke, JGC; Yubao, L; Jansen, JA. The influence of discharge power and heat treatment on calcium phosphate coatings prepared by RF magnetron sputtering deposition; *J. Mater. Sci.: Mater. Med.*, 18, (2007), 1061-1069.

[44] Kohale, RL; Dhoble, SJ. Synthesis and influence of Ce^{3+} codopant on the luminescence sensitivity of $SrZnP_2O_7:Dy^{3+}$ phosphor; *J. Alloys Compd.*, 586, (2014), 314-318.

[45] Yadav, RS; Rai, SB. Effect of concentration and wavelength on frequency downshifting photoluminescence from a Tb^{3+} doped yttria nano-phosphor: A photochromic phosphor; *J. Phys. Chem. Solids*, 114, (2018), 179-186.

[46] Yadav, RS; Rai, SB. Surface analysis and enhanced photoluminescence via Bi^{3+} doping in a Tb^{3+} doped Y_2O_3 nano-phosphor under UV excitation; *J. Alloys Compds.*, 700, (2017), 228-237.

[47] Monika, RS; Yadav, A; Bahadur, SB; Rai. Concentration and pump power-mediated color tunability, optical heating and temperature sensing via TCLs of red emission in an $Er^{3+}/Yb^{3+}/Li^+$ co-doped $ZnGa_2O_4$ phosphor doping; *RSC Adv.*, 9, (2019), 40092-40108.

[48] Yadav, RS; Dhoble, SJ; Rai, SB. Improved photon upconversion photoluminescence and intrinsic optical bistability from a rare earth co-doped lanthanum oxide phosphor via Bi^{3+} doping; *New J. Chem.*, 42, (2018), 7272-7282.

[49] Yadav, RV; Yadav, RS; Bahadur, A; Singh, AK; Rai, SB. Enhanced quantum cutting emission through Li^+ doping from Bi^{3+},

Yb^{3+} co-doped gadolinium tungstate phosphor; *Inorg. Chem.*, 55, (2016), 10928-10935.

[50] Yadav, RS; Kumar, D; Singh, AK; Rai, E; Rai, SB. Effect of Bi^{3+} ion on upconversion-based induced optical heating and temperature sensing characteristics in the Er^{3+}/Yb^{3+} co-doped La_2O_3 nanophosphor; *RSC Adv.*, 8, (2018), 34699-34711.

In: A Closer Look at Pyrophosphates
Editors: Ritesh L. Kohale et al.
ISBN: 978-1-53617-730-5
© 2020 Nova Science Publishers, Inc.

Chapter 3

OPTICAL PROPERTIES OF LANTHANIDE-BASED PYROPHOSPHATE PHOSPHOR MATERIALS

R. S. Yadav[1,*], Monika[2], S. J. Dhoble[3] and S. B. Rai[2]

[1]Department of Zoology, Institute of Science,
Banaras Hindu University, Varanasi, India
[2]Laser and Spectroscopy Laboratory, Department of Physics,
Institute of Science, Banaras Hindu University, Varanasi, India
[3]Department of Physics, R. T. M. Nagpur University, Nagpur, India

ABSTRACT

This chapter covers the synthesis, compositional effect and optical properties of lanthanide-based pyrophosphate phosphor materials. The general properties of the pyrophosphate materials have been discussed. The structural parameters of pyrophosphate phosphor materials have approved

---

[*] Corresponding Author's Email: ramsagaryadav@gmail.com.

that these materials are very promising and thermally stable. Additionally, the energy transfer between the lanthanide ions has been discussed by taking some examples. Basically, the two ions are involved in the energy transfer in which former one is called as donor ion whereas latter one is the accepter ion. The energy transfer takes place from donor to acceptor ions. This leads to an enhancement in the photoluminescence intensity of different phosphor materials. The addition of some surface modifiers also enhances the photoluminescence intensity of the phosphor materials. The applications of lanthanide-based pyrophosphate phosphor materials have been also incorporated in this chapter.

1. INTRODUCTION

Pyrophosphates are a member of phosphate family. They are phosphorus containing inorganic compounds in which two phosphorus atoms are attached in the P-O-P structure [1]. They are also known as diphosphates and can be obtained by neutralization of pyro-phosphoric acid. They are white or colorless compounds and the first member of polyphosphate series. The anionic form of pyrophosphate is $(P_2O_7)^{4-}$ ion and it is symbolically represented as:

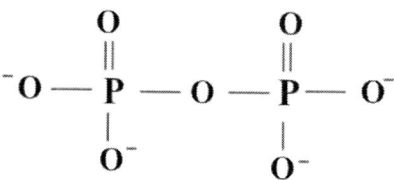

Figure 1. Chemical structure of pyrophosphate ion.

The structural and physical properties of pyrophosphates are extensively studied by various researchers. Dong et al. have prepared the $K_2Li_2P_2O_7$ diphosphates using high temperature solution method [2]. They have found that the $K_2Li_2P_2O_7$ diphosphates are well crystallized into the monoclinic phase with space group C2/c. The FTIR analysis indicates that the vibrational bands are present in the 427-583 cm^{-1} region and they are ascribed to the bending vibrations of δ(P–O–P) and δ(O–P–

O) molecules, respectively. This material is thermally stable as was confirmed by thermal gravimetric (TG) and differential scanning calorimetric (DSC) analyses. Three pyrophosphate $CsNaZnP_2O_7$, $RbNaZnP_2O_7$ and $RbLiMgP_2O_7$ materials were also synthesized by Song et al. using the same method [3]. The XRD patterns of all the pyrophosphates are crystalline in nature. In FTIR analysis, the bands at 930 and 720 cm^{-1} are also found in the three cases due to lattice vibrations of the P-O-P and O-P-O molecules, respectively. The thermal gravimetric (TG) and differential scanning calorimetric (DC) analyses further indicate high stability of these materials. The $Ba_2NaClP_2O_7$ pyrophosphate is also thermally stable material [4]. It shows a non-linear optical (NLO) behavior. White et al. have also prepared CeP_2O_7 material and found a phase transition from cubic to triclinic phases during compression at ~0.65 GPa and 5 GPa pressures [5].

The pyrophosphate is a low phonon energy host. The pyrophosphate based phosphors have attracted extensive interest due to a large number of applications in different fields, such as discharge tubes, fluorescent lamps, etc. [6-7]. Henderson et al. have prepared barium pyrophosphate phosphor doped with 30% titanium dioxide [6]. It emits an emission band covering the entire visible region with its peak centered at 483.5 nm. It also contains short afterglow with high efficiency and can be used in a discharge tube. A series of new pyrophosphate phosphors have been also prepared by Ranby et al. [7]. The concentration dependent emission intensity has also been studied by these authors. They have used Sn and Ce as primary activators and Mn as secondary activator in the alkaline earth pyrophosphate phosphors and reported their application in the fluorescent lamps.

The blue-green emitting pyrophosphate phosphor shows very interesting applications in solid state lighting and NUV light emitting diodes (LEDs) when Tb^{3+} doped $Sr_2P_2O_7$ phosphor is excited with near ultraviolet (NUV) radiation i.e., 352 nm [8]. It is a phosphor with mercury free excitation. On the other hand, the pyrophosphates are also useful in cathode materials for battery industry. Tanabe et al. have prepared sodium manganese pyrophosphate ($Na_2MnP_2O_7$) glass material by melt quenching method [9]. They have studied the crystallization behavior and electrochemical properties of the prepared glass and glass-ceramics. The crystallization of glass into glass ceramics occurs at 465°C. The $Na_2MnP_2O_7$ material also shows good electrochemical nature and can be used as cathode material in sodium ion based battery. Thus, the pyrophosphates are very thermally stable materials and have a large number of applications.

The incorporation of lanthanide ion (i.e., Eu^{3+} ion) was further explored in the $SrZnP_2O_7$ pyrophosphate phosphor. The phosphor emits intense red color due to $^5D_0 \rightarrow {}^7F_1$ transition (at 591 nm) on excitation with 254 nm [10]. Happek et al. have prepared Eu^{2+} doped $Cs_2M_2+P_2O_7$ (M = Ca, Sr) and observed orange and yellow photoluminescence emissions on excitation with 360 nm [11]. In another work, Eu^{2+} ion is further doped in the $SrMg_{1-x}Mn_xP_2O_7$ ($x = 0–1$) phosphors prepared through solid state reaction method [12]. The Eu^{2+} doped $SrMg_{1-x}Mn_xP_2O_7$ ($x = 0–1$) phosphors were excited by 350 nm radiation (i.e., x = 0.10, 0.15, 0.20, 0.30, 0.40, 0.50, 0.80 and 1.00). Due to overlapping between the emission of Eu^{2+} ion and excitation of Mn^{4+} ion there is possibility of energy transfer from Eu^{2+} to Mn^{4+} ions. This energy transfer enhanced the emission intensity of Mn^{4+} ion upto 120 times. They have also observed the color tunability in the phosphors with different doping concentrations of Mn^{4+} ions.

2. GENERAL PROPERTIES OF PYROPHOSPHATE

The pyrophosphates are white compounds and their alkali salts are easily soluble in water. They are known as pyrophosphates because they are obtained by heating the phosphates. The word 'pyro' is a Greek word and it is meant as heating. They can be treated as complexing agents for metal ions. The disodium pyrophosphates and tetrasodium pyrophosphates are very popular pyrophosphate compounds. As have been mentioned earlier that the pyrophosphate compounds are chemically, physically and thermally stable and have relatively low phonon energy. Therefore, they have been used as the host materials for doping various lanthanide and transition metal ions to get larger photoluminescence intensity from the phosphor materials.

The pyrophosphate materials have potential applications in different fields, such as solid state lighting, phosphor converted-LEDs, cathode materials for industry, etc. [8-12]. The pyrophosphate compounds also have numerous applications in the biological fields. When the nucleotide is added in the growing DNA/RNA by a polymerase the pyrophosphate is produced. It is also found in synovial fluid, urine and blood plasma in such an amount that it can prevent the calcination. Therefore, it also acts as inhibitor for calcination [13-14]. The β-calcium pyrophosphates are also used to improve the fusion success of posterolateral lumber fusion because it plays its role in bone graft extender [15].

3. Synthesis of Pyrophosphates

The pyrophosphate phosphor materials can be synthesized by different methods, such as solid state reaction method, modified solid state diffusion method, chemical precipitation method, etc [8, 16-18]. These methods produce the particles from micro-meter to nano-meter range. Aïcha Mbarek has synthesized a large number of pyrophosphate phosphors through solid state reaction method [16]. He has doped Eu^{3+} ion in the $ALnP_2O_7$ (where A = Rb, Cs, Tl and Ln = Y, Lu, Tm) pyrophosphates. The stoichiometric amounts of alkali metal carbonates, such as Cs_2CO_3, Rb_2CO_3, Tl_2CO_3 and rare earth oxides, such as Lu_2O_3, Tm_2O_3, Y_2O_3, etc. were weighed alongwith ammonium hydrogen phosphate $(NH_4)H_2PO_4$ compounds. These materials were ground properly in order to mix them in the homogeneous form. The obtained mixtures of different sets were heated initially at 400°C for 4 h and finally calcined them at 720°C for 20 h. He has further prepared the samples according to $ALn_{0.98}Eu_{0.02}P_2O_7$ compositions (where A = Cs, Rb, Tl and Ln = Lu, Tm, Y) using the same procedure, by substituting Eu^{3+} ions for Lu^{3+}, Tm^{3+} or Y^{3+} ions. These phosphors were used for structural and optical measurements.

Kohale et al. have prepared Tb^{3+} doped $Sr_2P_2O_7$ phosphor through high temperature modified solid state diffusion method [8]. In this method, they have used starting materials with high purity i.e., strontium carbonate ($SrCO_3$ with 99.995%), ammonium dihydrogen phosphate ($NH_4H_2PO_4$ with 99.99%) and terbium oxide (Tb_2O_3 with 99.99%). They have mixed the starting materials as discussed in the earlier case except the heating temperature. The samples were pre-heated at 350°C for 1 h and finally at 700°C for 12 h. In this method, large amount of gases were liberated, such as NH_3, CO_2, etc. as shown by the following reaction:

$$Sr(CO_3) + 2(NH_4H_2PO_4) \rightarrow Sr_2P_2O_7 + 2NH_3\uparrow + 3CO_2\uparrow + 3H_2O$$

They have also prepared the samples with different concentrations of Tb^{3+} ions using the same procedure. The obtained samples are white in color and they have been characterized for structural and optical properties.

The pyrophosphates were also prepared by chemical precipitation method. Recently, Wang et al. have prepared cobalt pyrophosphates ($Co_2P_2O_7$) by chemical precipitation method [18]. In this method, the analytical amounts of starting materials were collected in Teflon-lined stainless steel autoclave. The autoclave was sealed and run at 200°C for 12 h. After this, the white precipitate was obtained and it was washed many times by ethanol and distilled water. The product was dried in the air. Finally, the obtained powder was calcined at 450°C for 1 h.

4. OPTICAL PROPERTIES OF PYROPHOSPHATE PHOSPHORS

Basically, we discuss downconversion photoluminescence arising in the rare earth ions. Downconversion is a process in which the emitted photon has longer wavelength (i.e., lower energy) than the incident photon. The rare earth doped pyrophosphate phosphors have very interesting photoluminescence properties. In these phosphors, the photoluminescence occurs from the rare earth ions due to 4f-4f electronic transitions [1, 8, 16]. These transitions possess unique narrow band emissions. They are very intense and sharp. Usually, the energy levels of the rare earth ions are very dense and have long lifetimes. The presence of large number of energy levels offers an opportunity to play with a wide range of excitation wavelengths. One can excite a particular energy level corresponding to the interest of study. The rare earth ions emit light of different colors. It may be blue, green and green (RGB) or complementary colors [19-29]. The excitation and emission processes in the case of downconversion photoluminescence are shown in Figure 2. Figure 2(a) shows downconversion process in the rare earth ions. The ions present in ground state are irradiated by hv photon, which promotes them in the excited state. In the excited states, the relaxation of the ions from higher to lower excited states is termed as non-radiative transition. This leads to the population of the ions in the lower excited state. The transition from lower excited state to ground state is known as radiative transition. This gives visible photoluminescence and the emitted photon has lower energy compared to the incident photon.

On the other hand, the visible photoluminescence also occurs due to energy transfer between donor and acceptor ions. The process of energy transfer between donor and acceptor ions is shown in Figure 2(b). The energy transfer can be resonant or non-resonant. If the energy level of donor ion matches with the energy level of acceptor ion there is a resonant energy transfer. If it does not match to each other, the non-resonant energy transfer takes place. This can be possible with the coupling of phonon energy of the host lattice. In the resonant energy transfer, the emitted photon has the same energy to the incident photon. However, the emitted photon has lower energy compared to the incident photon in the non-resonant energy transfer process. This process can greatly enhance the photoluminescence intensity of the acceptor ion.

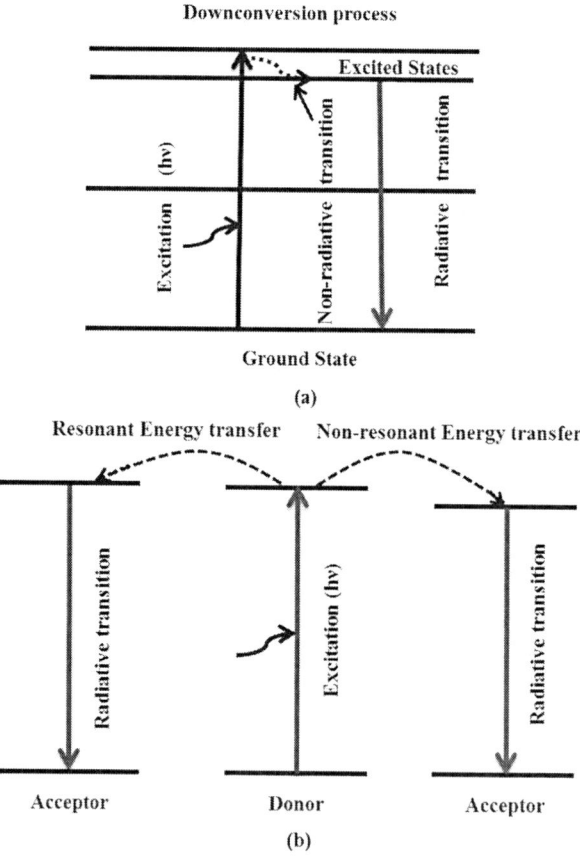

Figure 2. (a) Downconversion process and (b) resonant and non-resonant energy transfer processes between donor and acceptor ions.

Downconversion photoluminescence has been studied by Prasad et al. in the Sm^{3+} and Eu^{3+} doped calcium borophosphate ($2CaO$-B_2O_3-P_2O_5) phosphors [30]. They have prepared the Sm^{3+} and Eu^{3+} doped phosphors through solid state reaction method. The phosphors were prepared using different concentrations of dopant ions i.e., x = 0.2, 0.4, 0.6, 0.8 and 1.0 mol %. The concentrations of Sm^{3+} and Eu^{3+} ions were varied to see their effect on the photoluminescence intensity. They have excited the Sm^{3+} doped phosphors with 403 nm radiation whereas the Eu^{3+} doped phosphors were excited by using 394 nm wavelength. The phosphor sample emits orange red and red photoluminescence emission due to the $^4G_{5/2} \rightarrow {}^6H_{7/2}$ and $^5D_0 \rightarrow {}^7F_2$ transitions of Sm^{3+} and Eu^{3+} ions, respectively [27, 31]. It has been noticed that the photoluminescence intensity increases on increasing the concentrations of Sm^{3+} and Eu^{3+} ions. The photoluminescence intensity is optimum for 0.6 mol% concentrations in both the cases. The intensity of the samples decreases for higher concentrations of the dopant ions, which is due to concentration quenching effect. The CIE diagram confirms the presence of emitted colors in the orange red and red regions of visible spectrum.

Similar investigation was carried out by Wani et al. using Sm^{3+} and Eu^{3+} ions in the dilithium barium pyrophosphate ($Li_2BaP_2O_7$) phosphors [32]. They have also prepared the phosphor samples by solid state reaction method and varied the concentrations of Sm^{3+} and Eu^{3+} ions. The excitation spectra of Sm^{3+} and Eu^{3+} doped $Li_2BaP_2O_7$ phosphors reveal the intense peaks at 404 and 396 nm, respectively. The Sm^{3+} and Eu^{3+} doped $Li_2BaP_2O_7$ phosphors were irradiated by 404 and 396 nm excitations, which result reddish orange and red photoluminescence emissions due to the $^4G_{5/2} \rightarrow {}^6H_{7/2}$ and $^5D_0 \rightarrow {}^7F_1$ transitions of the Sm^{3+} and Eu^{3+} ions, respectively. The photoluminescence intensity is further reduced for higher concentrations due to concentration quenching. The critical distance between the ions has been calculated to explain the quenching process and it is found to be 3.85 nm for 0.5 mol% concentrations of Sm^{3+} ions. The CIE diagram further verifies the nature of the emitted colors. The Sm^{3+} and Eu^{3+} doped $Li_2BaP_2O_7$ phosphors may be useful in solid state lighting devices.

There is a possibility of energy transfer between Sm^{3+} and Eu^{3+} ions. The energy levels of Sm^{3+} and Eu^{3+} ions are distributed in such a way that the Sm^{3+} ions transfer their energy to Eu^{3+} ions. The energy transfer between the Sm^{3+} and Eu^{3+} ions has been studied by our group in the $La(OH)_3$ phosphor [33]. The orange red and red emissions are observed due to the $^4G_{5/2} \rightarrow {}^6H_{7/2}$ and $^5D_0 \rightarrow {}^7F_2$ transitions of Sm^{3+} and Eu^{3+} ions, respectively. When the Sm^{3+} and Eu^{3+} ions are co-doped together in the $La(OH)_3$ phosphor the emission intensity of red band of Eu^{3+} ion is increased due to energy transfer from Sm^{3+} to Eu^{3+} ions. Actually, the $^4G_{5/2}$ level of Sm^{3+} ion lies slightly above to that of the $^5D_0$ level of Eu^{3+} ion. This leads to an energy transfer from the Sm^{3+} and Eu^{3+} ions, which enhances the emission intensity of red band of Eu^{3+} ion. In this case, the Sm^{3+} is a donor ion whereas the Eu^{3+} acts as acceptor ion. The energy transfer is also verified by lifetime measurement. The lifetime of $^5D_0$ level of Eu^{3+} ion in the Eu^{3+} doped $La(OH)_3$ phosphor is 478 µs. Its value is increased to 691 µs due to presence of Sm^{3+} ion in the Eu^{3+} co-doped $La(OH)_3$ phosphor. There is another example of energy transfer reported using rare earth ions i.e., Dy^{3+}/Eu^{3+} ions in the $NaLa(MoO_4)_2$ phosphors [34]. Thus, the energy transfer improves the emission intensity of the phosphor materials.

The energy transfer has also been observed between transition metal ion and rare earth ion. Following this, we have studied the energy transfer from Bi^{3+} to Tb^{3+} ions in the Y_2O_3 phosphor prepared by using solution combustion method [22]. Both the ions give their characteristic emissions in the visible region. The energy transfer from Bi^{3+} to Tb^{3+} ions is shown in Figure 3. It shows that the Bi^{3+} and Tb^{3+} ions are excited by 266 nm in the Y_2O_3 phosphor and they emit blue, green, red and NIR emissions. The energy levels ($^1P_1$ and $^3P_1$) of Bi^{3+} ions are situated higher than that of the $^5D_4$ level of Tb^{3+} ions. The energy absorbed by Bi^{3+} ions in the energy levels ($^1P_1$ and $^3P_1$) is transferred to the $^5D_4$ level of Tb^{3+} ions. As a result, the emission intensity of different emission bands of

Tb^{3+} ions was enhanced significantly due to an energy transfer from Bi^{3+} to Tb^{3+} ions.

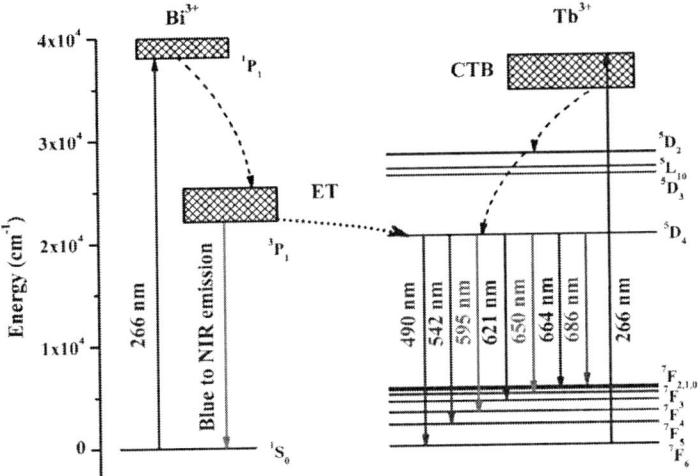

Figure 3. Energy level diagram of Bi^{3+} and Tb^{3+} ions alongwith energy transfer between Bi^{3+} to Tb^{3+} ions. (Reproduced with copyright permission from Elsevier, 2017).

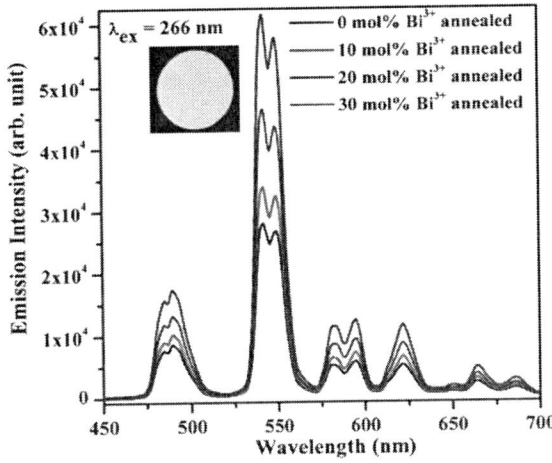

Figure 4. Effect of Bi^{3+} ions on the emission intensity of the Tb^{3+} doped Y$_2$O$_3$ phosphor upon 266 nm excitations and the inset image is color produced by the Bi^{3+}, Tb^{3+} co-doped Y$_2$O$_3$ phosphor. (Reproduced with copyright permission from Elsevier, 2017).

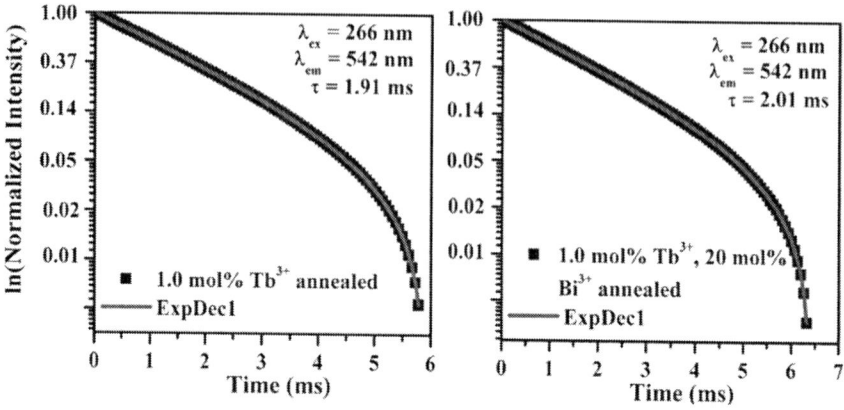

Figure 5. Decay curves for $^5D_4 \rightarrow {}^7F_5$ transition at 542 nm on excitation with 266 nm in the Bi^{3+}, Tb^{3+} co-doped Y_2O_3 phosphor. (Reproduced with copyright permission from Elsevier, 2017).

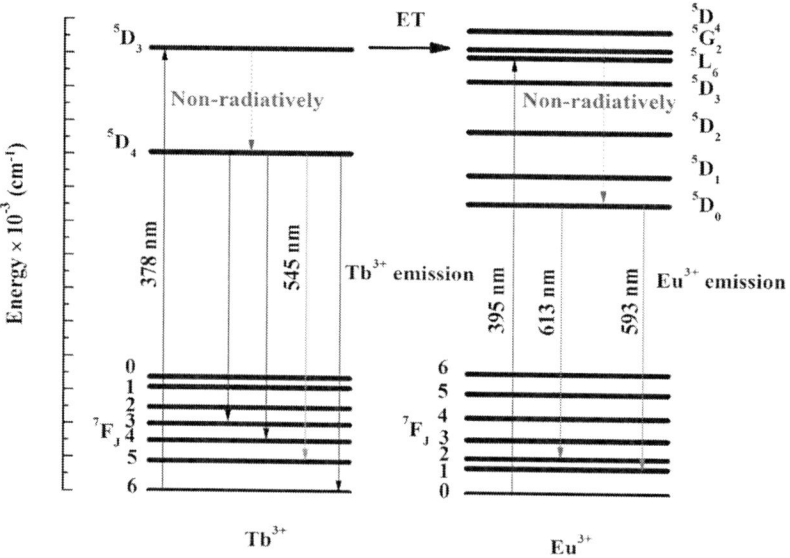

Figure 6. Energy levels of Tb^{3+} and Eu^{3+} ions, energy transfer from Tb^{3+} and Eu^{3+} ions and different electronic transitions. (Reproduced with copyright permission from Royal Society of Chemistry, 2017).

The effect of Bi^{3+} ions on the emission intensity of the Tb^{3+} doped Y_2O_3 phosphor has been also investigated. The presence of Bi^{3+} ion improves the crystallinity and particles size of the Tb^{3+} doped Y_2O_3

phosphor. In this case, it acts as surface modifier. When Bi^{3+} ion transfers its energy to Tb^{3+} ions it behaves as sensitizer ion. Thus, the Bi^{3+} ion improves the emission intensity of the Tb^{3+} doped Y_2O_3 phosphor due to these combined parameters i.e., improvement in crystallinity, particles size and energy transfer. The effect of Bi^{3+} ions on the emission intensity of the Tb^{3+} doped Y_2O_3 phosphor is shown in Figure 4. The presence of Bi^{3+} ion increases the emission intensity upto two times.

The energy transfer between Bi^{3+} to Tb^{3+} ions can also be understood by monitoring the lifetimes of the emitting level. The lifetime of the $^5D_4$ level of Tb^{3+} ion was calculated for $^5D_4 \rightarrow {}^7F_5$ transition at 542 nm on excitation with 266 nm. It is found to be 1.91 ms. When Bi^{3+} is doped in the $Tb^{3+}:Y_2O_3$ phosphor the lifetime of the $^5D_4$ level of Tb^{3+} ion was increased to 2.01 ms. This is due to an increase in the emission intensity due to energy transfer process. The decay curves thus obtained in both the cases are shown in Figure 5. The energy transfer process has been also observed in the other combination of lanthanide and transition metal ions in the pyrophosphate phosphors i.e., Ce^{3+}, Mn^{2+}-doped $K_2AEP_2O_7$ (where AE = Ca, Sr) phosphor [35]. The energy transfer from Ce^{3+} to Mn^{2+} ions has been found to enhance the intensity of the emission band of Mn^{2+} ion.

There is another example of energy transfer between Tb^{3+} and Eu^{3+} ions in the $Ba_2P_2O_7$ phosphor. The Tb^{3+} and Eu^{3+} doped and co-doped $Ba_2P_2O_7$ phosphors were prepared by solid state reaction method [36]. The Tb^{3+} and Eu^{3+} doped $Ba_2P_2O_7$ phosphors were excited using 378 and 395 nm wavelengths, respectively corresponding to intense peaks in their excitation spectrum. The Tb^{3+} doped phosphor gives green emission ($^5D_4 \rightarrow {}^7F_5$ transition) whereas Eu^{3+} doped $Ba_2P_2O_7$ phosphor emits red emission ($^5D_0 \rightarrow {}^7F_1$ transition), respectively. When the Tb^{3+}, Eu^{3+} co-doped phosphor is excited with 378 nm, the intensity of red emission of Eu^{3+} ion increases whereas that of green emission of Tb^{3+} ion is decreased. This is due to an energy transfer from Tb^{3+} to Eu^{3+} ions due to dipole-dipole interactions. The energy transfer from Tb^{3+} to Eu^{3+} ions is shown in Figure 6. The energy transfer takes place from $^5D_3$ level of

Tb^{3+} ions to $^5G_2$ level of Eu^{3+} ions. They have also studied the effect of alkali ions (i.e., Li$^+$, Na$^+$, K$^+$) on the emission intensity of the Eu^{3+} co-doped Ba$_2$P$_2$O$_7$ phosphor. It was found that the emission intensity of Eu^{3+} bands is enhanced in the presence of alkali ions; however, the emission intensity is larger of Na$^+$ doping. The improvement in the emission intensity is due to charge compensation in the Ba$_2$P$_2$O$_7$ phosphor generated by alkali ions. In this case, the Tb^{3+} ion acts as sensitizer whereas the alkali ion plays the role of a surface modifier.

5. APPLICATIONS OF PYROPHOSPHATE PHOSPHORS

The pyrophosphate phosphors are also prepared by using transition metal ions. The Mn^{4+} is a transition metal ion and it is doped in the potassium gallium pyrophosphate (KGaP$_2$O$_7$) host material [17]. Naresh et al. have also prepared the phosphor samples by solid state reaction method using different concentrations of Mn^{4+} ions. They have studied the structural, optical and temperature dependent properties of these phosphors. They have monitored the UV-Vis diffuse reflectance spectra of the phosphors and observed a charge transfer transition in the 225-300 nm due to O$^{2+} \rightarrow$ Mn^{4+}, a weak band at 332 nm due to $^4A_{2g} \rightarrow {}^4T_{1g}$ and an intense broad band at 453 nm due to $^4A_{2g} \rightarrow {}^4T_{2g}$ transition of Mn^{4+} ions, respectively. The excitation spectra also have two intense bands at 338 and 452 nm, which are further assigned due to $^4A_{2g} \rightarrow {}^4T_{1g}$ and $^4A_{2g} \rightarrow {}^4T_{2g}$ transitions of Mn^{4+} ions, respectively. Since the excitation peak at 452 nm is intense enough, therefore, it has been used for exciting the samples. The phosphor samples were prepared with different concentrations of Mn^{4+} ions i.e., 0.01, 0.03, 0.05, 0.07, 0.09 and 0.11 mol%. The Mn^{4+} doped phosphor emits strong red photoluminescence at 691, 693 and 696 nm wavelengths due to $^2E_g \rightarrow {}^4A_{2g}$ transitions of Mn^{4+} ions. The temperature dependent photoluminescence shows a regular decrease in the intensity of red band due to increased lattice vibrations of the host lattices. They have suggested the applications of Mn^{4+} doped $KGaP_2O_7$ phosphor in the field of red emitting light emitting diodes, horticulture and photonic devices. The Mn^{2+} co-doped $(Sr,Sn)ZnP_2O_7$ phosphor also gives emissions in the blue-green to red regions with increasing concentrations of Mn^{2+} ion, which is due to an energy transfer from Sn^{2+} to Mn^{2+} ions [37]. The change in the emitted color leads to generate white light. Therefore, this phosphor is expected to be used for white light emitting sources.

Another rare earth free phosphor was also found to give blue photoluminescence when Co^{2+} is doped in the $K_2ZnP_2O_7$ phosphor [38]. The phosphor sample was excited by 272 nm and emission was observed in the near blue region, which is centered at 386 nm. The color emitted by the phosphor was confirmed by CIE diagram. The correlated color temperature was calculated using McCamy polynomial formula for (0.5, 1.0, 1.25 and 1.50 mol%) and their values are found to be 2282, 1750, 2250 and 1884, respectively. These values are in well agreement with the warm light. This phosphor finds its application in warm white light LEDs.

The rare earth doped phosphor materials are of great scientific importance and have wide applications in different fields. The Eu^{3+} doped $AgLaP_2O_7$ phosphor was prepared by Hami et al. using solid state reaction and they have used spectroscopic tools to study the optical properties of the phosphor [39]. The TGA/DTA analyses show that this pyrophosphate phosphor is highly stable. The Eu^{3+} doped $AgLaP_2O_7$ phosphor emits red color due to $^5D_0 \rightarrow {}^7F_2$ transition of Eu^{3+} ion upon 395 nm radiations. This shows that the phosphor can be easily excited using near ultraviolet (n-UV) light. Hakeem et al. have also prepared $Ca_{1-x}MgP_2O_7:xEu^{3+}$ pyrophosphate phosphor, where x = 0.01, 0.03, 0.05, 0.10 and 0.20 mol% and studied the photoluminescence properties using 393 nm excitations [40]. The phosphor sample further gives strong red photoluminescence at 0.10 mol% concentration of Eu^{3+} ion due to $^5D_0 \rightarrow {}^7F_2$ transition. The FTIR analysis of the phosphor reveals that the phosphor contains low phonon energy, which is essential for large photoluminescence intensity. This phosphor can be used as a red emitting phosphor for white LEDs.

The pyrophosphate phosphors also emit blue and green light depending on the selection of rare earth ions. The Ce^{3+} doped dicalcium pyrophosphoate ($Ca_2P_2O_7$) phosphor emits near blue emissions at 362 and 390 nm due to 5d→4f transition of Ce^{3+} ion when it is excited by 298 nm radiations. The Ce^{3+} doped $Ca_2P_2O_7$ phosphor may be used in the scintillation purposes [41]. In a similar work, the Ce^{3+} ion is further doped in the same phosphor and prepared at different annealing temperatures. The pure sample was identified as β-$Ca_2P_2O_7$ phosphor, which has been confirmed by XRD analysis. This phosphor gives thermoluminescence (TL) glow and can be used in radiation dosimetry purposes [42]. In another case, the Eu^{2+} and Ce^{3+} ions were separately doped in the $SrCaP_2O_7$ phosphor. The Eu^{2+} and Ce^{3+} doped $SrCaP_2O_7$ phosphors were excited by 330 and 300 nm wavelengths and they emit blue (at 425 nm) and near UV (at 337 nm) emissions, respectively [43]. Since the Eu^{2+} doped $SrCaP_2O_7$ phosphor is excited with n-UV light, therefore, this can be treated as blue emitting candidate for white LEDs. On the other hand, the Ce^{3+} doped $SrCaP_2O_7$ phosphor can be used in lamp industry and scintillation purposes. Kohale et al. have discussed the application of Eu^{2+} doped $SrCaP_2O_7$ phosphor in the blue emitting solid state lighting purposes [44]. They have also found blue emission at 427 nm upon 330 nm excitation. The Eu^{2+} doped $SrCaP_2O_7$ hafnium pyrophosphate phosphor is also a source of blue emitting material, which yields blue emission at 450 nm excited by 330 nm radiations [45].

Ma et al. have prepared the Ce^{3+} and Tb^{3+} doped $Sr_2(P_2O_7)$ phosphors by co-precipitation method [46]. The prepared phosphors are highly crystalline. The phosphor emits intense green emission at 542 nm due to $^5D_4 \rightarrow {^7F_5}$ transition of Tb^{3+} ion on excitation with 254 nm. The photoluminescence intensity is larger for the annealing time for 4 h compared to 2 and 3 h. The CIE coordinates indicate that this phosphor is a green emitting candidate for UV-converted white light sources. As were mentioned earlier that the pyrophosphates also have a large number of applications in biological fields. They can be used in the DNA/RNA processing, inhibitor for calcination, bone graft extender, etc. [13-15]. It has a tendency to detect various microbial activities. The alkaline pyrophosphate plays its role in understanding various processes of physiochemical and pathological behaviors. The inorganic pyrophosphate plays the role of metabolite for biochemical processes, such as arthritis, chondrocalcinosis, etc. [47-48].

CONCLUSION

In this chapter, the general properties of the pyrophosphate materials have been discussed. The structural and physical properties of the pyrophosphate materials suggest that they are high thermally stable materials and have low phonon energy. Different types of pyrophosphate materials were incorporated into this chapter. Some synthesis procedures were also mentioned for preparing the pyrophosphate materials. The basis properties of the rare earth ions have also been discussed. The energy transfer between the lanthanide and transition metal ions has been discussed by using the schematic energy level diagram. The addition of some sensitizers and surface modifiers were found to enhance the photoluminescence intensity of the phosphor materials. The optical properties of pyrophosphate materials were discussed not only in lanthanide ions but also in transition metal ions. They have contributed their role in design and fabrication of solid state lighting devices, green emitting sources, phosphor converted-white LEDs, scintillation, radiation dosimetry, lamp industry, etc. The pyrophosphate phosphors materials have also been found their applications in physiochemical and pathological behaviors of biological systems, arthritis, chondrocalcinosis, etc.

REFERENCES

[1] Wani, J. A., N. S. Dhoble, S. P. Lochab, S. J. Dhoble, Luminescence characteristics of C^{5+} ions and ^{60}Co irradiated $Li_2BaP_2O_7:Dy^{3+}$ phosphor, *Nuclear Inst. Meth. Phys. Res. B* 349 (2015) 56-63.

[2] Dong, L., X. Ge, P. Zhang, Q. Li, S. Sun, S. Shen, Further example of diphosphates: synthesis and characterization of $K_2Li_2P_2O_7$, *Z. Anorg. Allg. Chem.* 645 (2019) 944-948.

[3] Song, H., S. Zhang, Y. Li, W. Liu, Z. Lin, J. Yao, G. Zhang, Syntheses, crystal structures, and characterizations of three new pyrophosphates $CsNaZnP_2O_7$, $RbNaZnP_2O_7$, and $RbLiMgP_2O_7$, *Solid State Sci.* 95 (2019) 105940 pp 1-7.

[4] Chen, J., L. Xiong, L. Chen, L. M. Wu, $Ba_2NaClP_2O_7$: Unprecedented phase matchability induced by symmetry breaking and its unique fresnoite-type structure *J. Am. Chem. Soc.*, 2018 (140) 14082-14086.

[5] White, K. M., P. L. Lee, P. J. Chupas, K. W. Chapman, E. A. Payzant, A. C. Jupe, W. A. Bassett, C.-S. Zha, A. P. Wilkinson, Synthesis, symmetry, and physical properties of cerium pyrophosphate, *Chem. Mater.* 20 (2008) 3728-3734.

[6] Henderson, S. T., P. W. Ranby, Barium titanium phosphate: A new phosphor, *J. Electrochem. Soc.*, 98 (1951) 479-482.

[7] Ranby, P. W., D. H. Mash, S. T. Henderson, The investigation of new phosphors, with particular reference to the pyrophosphates, *Br. J. Appl. Phys.* 6 (1955) S18-S25.

[8] Kohale, R. L., O. P. Chimankar, S. J. Dhoble, Blue-green luminescence in Hg free excited $Sr_2P_2O_7:Tb^{3+}$, pyrophosphate phosphor for NUV excited LEDs, *IOP Conf. Ser.: Mater. Sci. Eng.* 73 (2015) 012111 pp1-4.

[9] Tanabe, M., T. Honma, T. Komatsu, Unique crystallization behavior of sodium manganese pyrophosphate $Na_2MnP_2O_7$ glass

and its electrochemical properties, *J. Asian Cer. Soc.* 5 (2017) 209-215.

[10] Qin, L., C. Xu, Y. Huang, S. Kim, H. J. Seo, Spectroscopic and structural features of Eu^{3+}-doped zinc pyrophosphate ceramic, *Cer. Int.* 40 (2014) 1605-1611.

[11] Happek, U., St. Chaney, M. Aycibin, A. Srivastava, The luminescence of octahedrally coordinated divalent europium in $Cs_2M_2 + P_2O_7$ (M = Ca, Sr), *ECS Trans.* 16 (2009) 57-60.

[12] Li, K., D. Chen, R. Zhang, Y. Yu, Y. Wang, $Eu^{2+}:SrMg_{1-x}Mn_xP_2O_7$ (x = 0-1) phosphors with tunable yellow-red emissions, *J. Alloys Compds.* 555 (2013) 45-50.

[13] Ho, A. M., M. D. Johnson, D. M. Kingsley. Role of the mouse ank gene in control of tissue calcification and arthritis, *Sci.* 289 (2000) 265-270.

[14] Lahti, R., Microbial inorganic pyrophosphatases, *Microbio. Rev.* 47 (1983) 169-179.

[15] Lee, J. H., B. S. Chang, U. O. Jeung, K. W. Park, M. S. Kim, C.-K. Lee, The first clinical trial of beta-calcium pyrophosphate as a novel bone graft extender in instrumented posterolateral lumbar fusion, *Clinics Orthop. Surg.* 3 (2011) 238-244.

[16] Mbarek, A., Synthesis, structural and optical properties of Eu^{3+}-doped $ALnP_2O_7$ (A = Rb, Cs, Tl; Ln = Y, Lu, Tm) pyrophosphates phosphors for solid-state lighting, *J. Mol. Structure* 1138 (2017) 149-154.

[17] Naresh, V., Nohyun Lee, $KGaP_2O_7: Mn^{4+}$ deep red emitting phosphor: Synthesis, structure, concentration and temperature dependent photoluminescence concentration and temperature dependent photoluminescence, *J. Lumin.* 214 (2019) 116565 pp 1-11.

[18] Wang, W. P., H. Pang, M. L. Jin, X. Shen, Y. Yao, Y. G. Wang, Y. C. Li, X. D. Li, C. Q. Jin, R. C. Yu, Studies on the structural stability of $Co_2P_2O_7$ under pressure, *J. Phys. Chem. Solids* 116 (2018) 113-117.

[19] Yadav, R. S., R. K. Verma, S. B. Rai, Intense white light emission in $Tm^{3+}/Er^{3+}/Yb^{3+}$ co-doped Y_2O_3-ZnO nano-composite, *J. Phys. D: Appl. Phys.* 46 (2013) 275101 pp 1-8.

[20] Yadav, R. S., S. B. Rai, Structural analysis and enhanced photoluminescence via host sensitization from a lanthanide doped $BiVO_4$ nano-phosphor, *J. Phys. Chem. Solids* 110 (2017) 211-217.

[21] Yadav, R. S., S. B. Rai, Frequency upconversion and downshifting emissions from rare earth co-doped strontium aluminate nano-phosphor: A multi-modal phosphor, *J. Lumin.* 190 (2017) 171-178.

[22] Yadav, R. S., S. B. Rai, Surface analysis and enhanced photoluminescence via Bi^{3+} doping in a Tb^{3+} doped Y_2O_3 nano-phosphor under UV excitation, *J. Alloys Compds.* 700 (2017) 228-237.

[23] Yadav, R. S., S. B. Rai, Effect of concentration and wavelength on frequency downshifting photoluminescence from a Tb^{3+} doped yttria nano-phosphor: A photochromic phosphor, *J. Phys. Chem. Solids* 114 (2018) 179-186.

[24] Yadav, R. S., D. Kumar, A. K. Singh, E. Rai, S. B. Rai, Effect of Bi^{3+} ion on upconversion-based induced optical heating and temperature sensing characteristics in the Er^{3+}/Yb^{3+} co-doped La_2O_3 nano-phosphor, *RSC Adv.* 8 (2018) 34699-34711.

[25] Yadav, R. S., S. J. Dhoble, S. B. Rai, Enhanced photoluminescence in Tm^{3+}, Yb^{3+}, Mg^{2+} tri-doped $ZnWO_4$ phosphor: Three photon upconversion, laser induced optical heating and temperature sensing, *Sens. Actua. B: Chem.* 273 (2018) 1425-1434.

[26] Yadav, R. S., S. J. Dhoble, S. B. Rai, Improved photon upconversion photoluminescence and intrinsic optical bistability from a rare earth co-doped lanthanum oxide phosphor via Bi^{3+} doping, *New J. Chem.* 42 (2018) 7272-7282.

[27] Yadav, R. S., S. B. Rai, Effect of annealing and excitation wavelength on the downconversion photoluminescence of Sm^{3+} doped Y_2O_3 nano-crystalline phosphor, *Opt. Laser Technol.* 111 (2019) 169-175.

[28] Baig, N., R. S. Yadav, N. S. Dhoble, V. L. Barai, and S. J. Dhoble, Near UV excited multi-color photoluminescence in RE^{3+} (RE = Tb, Sm, Dy and Eu) doped $Ca_2Pb_3(PO_4)_3Cl$ phosphors, *J. Lumin.* 215 (2019) 116645 pp 1-8.

[29] Gu, F., C. Li, H. Cao, W. Shao, Y. Hu, J. Chen et al. Crystallinity of Li-doped MgO: Dy^{3+} nanocrystals via combustion process and their photoluminescence properties, *J. Alloys Compds.*, 453 (2008) 361-365.

[30] Prasad, V. R., S. Damodaraiah, S. Babu, Y. C. Ratnakaram, Structural, optical and luminescence properties of Sm^{3+} and Eu^{3+} doped calcium borophosphate phosphors for reddish-orange and red emitting light applications, *J. Lumin.* 187 (2017) 360-367.

[31] Yadav, R. V., R. S. Yadav, A. Bahadur, S. B. Rai, Down shifting and quantum cutting from Eu^{3+}, Yb^{3+} co-doped $Ca_{12}Al_{14}O_{33}$ phosphor: A dual mode emitting material, *RSC Adv.* 6 (2016) 9049-9056.

[32] Wani, J. A., N. S. Dhoble, N. S. Kokode, B. D. P. Raju, S. J. Dhoble, Synthesis and luminescence property of $Li_2BaP_2O_7$: Ln^{3+} (Ln = Eu, Sm) phosphors, *J. Lumin.* 147 (2014) 223-228.

[33] Yadav, R. S., Y. Dwivedi, S. B. Rai, Structural and optical properties of Eu^{3+}, Sm^{3+} co-doped $La(OH)_3$ nano-crystalline red emitting phosphor, Spectrochim. *Acta Part A* 132 (2014) 599-603.

[34] Du, P., J. S. Yu, Energy transfer mechanism and color controllable luminescence in Dy^{3+}/Eu^{3+} co-doped $NaLa(MoO_4)_2$ phosphors, *J. Alloys Compds.* 653 (2015) 468-473.

[35] Hatwar, L. R., S. P. Wankhede, S. V. Moharil, P. L. Muthal, and S. M. Dhopte, Luminescence and energy transfer in Ce^{3+}, Mn^{2+}-doped $K_2AEP_2O_7$ (AE = Ca, Sr) phosphor, *Lumin.* 30 (2015) 904-909.

[36] Wang, B., Q. Ren, O. Hai, X. Wu, Luminescence properties and energy transfer in Tb^{3+} and Eu^{3+} co-doped $Ba_2P_2O_7$ phosphors, *RSC Adv.* 7 (2017) 15222-15227.

[37] Kim, S. W., T. Hasegawa, H. Yumoto, T. Ishigaki, K. Uematsu, K. Toda, M. Sato, Synthesis and photoluminescence properties of

Mn^{2+} co-doped white emitting (Sr,Sn)ZnP$_2$O$_7$ phosphor, *J. Cer. Proc. Res.* 15 (2014) 177-180.

[38] Belbal, R., L. Gacem, B. Bentria, Blue emission of Co^{2+} in K$_2$ZnP$_2$O$_7$ phosphors, *Inorg. Chem. Comm.* 97 (2018) 39-43.

[39] Hami, W., D. Zambon, A. Zegzouti, M. Elaatmani, M. El-Ghozzi, M. Daoud, Application of spectroscopic properties of Eu^{3+} ion to predict the site symmetry of active ions in AgLaP$_2$O$_7$:Eu^{3+} phosphors, *Inorg. Chem. Comm.* 107 (2019) 107475 pp 1-9.

[40] Hakeem, A., K. N. Shinde, S. J. Yoon, S. J. Dhoble, K. Park, Synthesis and photoluminescence of Novel Ca$_{1-x}$MgP$_2$O$_7$:xEu^{3+} pyrophosphate phosphor for near ultraviolet light emitting diodes, *J. Nanosci. Nanotechnol.* 14 (2014) 5873-5876.

[41] Kohale, R. L., S. J. Dhoble, Optical performance of Ca$_2$P$_2$O$_7$:Ce^{3+} pyrophosphate phosphor synthesized via modified solid state diffusion, *J. Mol. Structure* 1170 (2018) 18-23.

[42] Lozano, I. B., J. Roman-Lopez, R. Sosa, J. A. IDíaz-Góngora, J. Azorín, Preparation of cerium doped calcium pyrophosphate: Study of luminescent behavior, *J. Lumin.* 173 (2016) 5-10.

[43] Shinde, K. N., Luminescence in Eu^{2+} and Ce^{3+} doped SrCaP$_2$O$_7$ phosphors, *Results Phys.* 7 (2017) 178-182.

[44] Kohale, R. L., S. J. Dhoble, Eu^{2+} luminescence in SrCaP$_2$O$_7$ pyrophosphate phosphor, *Lumin.* 28 (2013) 656-661.

[45] Kim, S. W., Y. Zuo, T. Masui, N. Imanaka, Novel blue-emitting phosphors based on hafnium pyrophosphate, *Mater. Lett.* 95 (2013) 97-99.

[46] Ma, C. G., W. Zheng, L. G. Jin, L. M. Dong. Fluorescence and preparation of Sr$_2$(P$_2$O$_7$):Ce,Tb phosphate by co-precipitation method, *Rare Met.* (2013) 32(4):420-424.

[47] Abhishek, A., M. Doherty, Pathophysiology of articular chondrocalcinosis-role of ANKH, *Nat. Rev. Rheumatol.* 7 (2011) 96-104.

[48] Wang, F., C. Zhang, Q. Xue, H. Li, Y. Xian, Label-free upconversion nanoparticles-based fluorescent probes for sequential

sensing of Cu^{2+}, pyrophosphate and alkaline phosphatase activity, *Biosens. Bioelect.* 95 (2017) 21-26.

In: A Closer Look at Pyrophosphates
Editors: Ritesh L. Kohale et al.
ISBN: 978-1-53617-730-5
© 2020 Nova Science Publishers, Inc.

Chapter 4

THE PHARMACOLOGY OF PYROPHOSPHORIC ACID: A CRITICAL ANALYSIS OF THE FIRST QUARTER CENTURY OF RESEARCH

William Banks Hinshaw[1,] and Allyn F. DeLong[2]*

[1]Department of Surgery, Harris Regional Hospital, Sylva, NC, US
[2]Indiana Environmental Protection Agency (retired), Carmel, IN, US

ABSTRACT

Studies involving the physical chemical and biochemical interaction of exogenous pyrophosphoric acid with bone became the focus of some of the research of four major teams during the interval around 1950-1975. There proved to be remarkable progress during this era from the naive assumption that the physical chemical interaction of pyrophosphoric acid with bone mineral was a controlling factor to the realization that the molecule shared specific regulatory properties in common and in

---

[*] Corresponding Author's Email: williambh@frontier.com.

conjunction with a number of recognized hormonal influences, but was limited to these functions in specific protected environments.

We summarize here the contributions of William F. Neuman, Herbert Fleisch, Hajimi Orimo, and Howard Rasmussen and their collaborators toward defining the influence of this small molecule in the process of bone remodeling as we understand it today.

These functions will be shown, in a subsequent chapter, to be remarkably parallel to some of the functions of some widely employed pharmacodynamic agents used today in the management of postmenopausal and other forms of bone loss.

ABBREVIATIONS

PPi   inorganic pyrophosphate;
Pi   inorganic orthophosphate;
TPTX   thyroidparathyroidectomized

INTRODUCTION

When the informed reader sees this chapter title, a doubt may arise as to its value, since pyrophosphoric acid (PPi) has little or no application in human or veterinary medicine. In fact, at pharmacologic doses, the compound is poorly tolerated. Some uses have been found as a descaler and a paper sealant and PPi has been used in the plumbing and packaging industries for these purposes, but the occasional use in toothpaste for descaling seems to have become unpopular. However, it is our privilege here to summarize the state of knowledge regarding the measurable effects of administration of exogenous PPi in animal systems, because a quite inaccurate opinion may be formed based on recent (Orriss, Arnett, and Russell 2016) reviews as well as earlier basic research publications (Fleisch et al. 1966; Reynolds et al. 1972). The early fundamental research (Rasmussen et al. 1970) of one of us (AFD) and the near-simultaneous findings in Japan (Orimo, Fujita, and Yoshikawa 1969) are inconsistent with the proposal (Orriss, Arnett, and Russell 2016) that this compound acts purely as an inhibitor of mineralization at pharmacologic doses. The actual dose, and the route of dosing, however, are the critical points that have been ignored by several research groups. The importance of this understanding is crucial to the argument presented in the following chapter.

Endogenous pyrophosphate[1] has vitally important functions in human and animal metabolism. Among other sources, it is the byproduct of the hydrolysis of triphosphate nucleosides in the formation and repair of the nucleic acids and is estimated to be produced daily in humans in gram quantities. Despite its evident toxicity in both health and disease, the molecule is subject, in the peripheral circulation, to enzymatic hydrolysis converting a sufficient percentage to relatively benign orthophosphate. PPi can survive *in vivo* in a protected sequestered environment where it plays a major role in the regulation of bone metabolism (Terkletaub 2001; Harmey et al. 2004). Assumptions made, that parenterally administered PPi undergoes rapid hydrolysis, must be contrasted with other evidence demonstrating that given intravenously, PPi is rapidly taken up into such a sequestered environment. These details will be fully explained in this review.

EARLIEST PUBLICATIONS RELATING TO PYROPHOSPHORIC ACID IN ANIMAL METABOLISM

When the distinguished American biochemist William F. Neuman first began to collaborate with his Swiss postdoctoral fellow Herbert Fleisch around 1959, it was well known that the extracellular fluid of organisms with endoskeletons contained sufficient calcium and

---

[1] Chemists may take exception to the term 'pyrophosphate' used ubiquitously in the biological literature for ionized pyrophosphoric acid, specifically because this acid is usually being considered as functioning at physiologic pH 7.4. However, this is the convention which shall be followed here and in the following chapter.

phosphorus that crystallization could occur if nucleation (seeding) was present. Animal collagen induced this crystallization under *in vitro* aqueous experimental conditions. This team surmised that some entity was present in living systems which prevented this crystallization, and that this entity must be removed in order for crystallization to begin. Neuman had postulated this "detoxicating" factor as early as 1950 (Neuman 1950) and by the time his collaboration with Fleisch began, also believed that this function was fulfilled by the enzyme alkaline phosphatase. Their experiments demonstrated that the Ca X Pi product (the concentration) required to induce crystallization in dog serum was reduced to that of water controls by the addition of alkaline phosphatase. The hunt was on from that point to discover the physiologic inhibitor present in the serum.

Pyrophosphate and several polyphosphates proved to be the most potent inhibitors of crystal initiation by collagen. In the last paper that they published together, Fleisch and Neuman (1961) still had not decided precisely what the inhibitor was, but PPi and the polyphosphates were considered the most likely. It remained for Fleisch, when he returned to Switzerland, to discover that dilute urine contained such an inhibitor. He was able to isolate and confirm that this substance was pyrophosphoric acid (Fleisch and Bisaz 1962).

LATER EXPERIMENTS IN DAVOS THAT ADDRESSED THE PROPERTIES OF PYROPHOSPHORIC ACID

By 1965, Fleisch had expanded his cadre of collaborators working in his Swiss laboratories in Davos. The interval up to that time was occupied with a number of inorganic chemistry experiments regarding the effect of PPi on calcium phosphate precipitation *in vitro*. Much earlier Fleisch and Neuman had expressed an understanding that calcium precipitation experiments in living tissue culture were preferable. Many of his publications describe experiments with multiple objectives. This review will be limited to those objectives that have some bearing on the properties of PPi. It seems ironic that one of his first experimental protocols at Davos revealed the earliest clue that pyrophosphate might have dual characteristics depending on the dose present, a finding that was never followed up at Davos.

Investigation of the effect of PPi in ex-vivo experiments using cultured living embryonic chick femurs revealed this duality. As could be anticipated from the preceding work, PPi (as well as an imperfectly characterized mixture of polyphosphates know as Graham salt) appeared to inhibit mineralization (determined by the mass difference from controls) of this growing tissue at pyro/polyphosphoric concentrations of 4 and 16 μgms P/ml. But at a concentration of 1 μgm P/ml both phosphorus entities increased net bone mineral "deposition" (Fleisch et al. 1966). The potency of this effect was high, despite being described in the abstract as "slight." The net difference in mass by two measurement techniques in the bone exposed to the 1μgm P/ml additive was +45% more than control. This should be compared with the "suppressive" effect of the next higher dose (4 μgm P/ml) which was negative: -26%. Thus by 1966, evidence had been acquired revealing a dual role for PPi in living tissue. This was an opportunity, rarely discernible from an analysis of published literature alone, to recognize the moment when a significant discovery was missed. In the paper (in contrast to the abstract) Fleisch wrote:

> "It is interesting that, at a concentration of 1 μgm P/ml, both compounds seemed to activate calcification, a property which was not observed in our *in vitro* studies. This suggests another facet of regulation of calcification by pyrophosphate. It inhibits at certain concentrations; its partial destruction by pyrophosphatase would then not only destroy this

inhibition but might also lead to facilitation of calcium phosphate deposition[2]. (Fleisch et al. 1966)"

From this explicit conclusion, it is clear that the authors were still thinking in terms of the physical chemical effect of the polyphosphate on the nucleation ability of the living tissue. This was soon to change.

Around this same time, Fleisch established a professional relationship with Graham Russell and co-published with him in the same year on work acknowledging Sandoz Stiftung zur Forderung der Medizinisch-Biologischen Wissenschaften financial support (Fleisch, Russell, and Strauman 1966a). At the end of the above pivotal chicken embryonic femur paper Fleisch wrote "We are grateful to… Mr. Graham Russell for helpful advice." From this point on, R. Graham G. Russell co-authored many papers with Herbert Fleisch and has continued and extended Fleisch's work up to the present day. In a recent review (Orriss, Arnett, and Russell 2016), Russell ignored even the tepid conclusion expressed in the embryonic chick femur paper abstract (certainly not tepid in the complete paper) regarding the "small activation of mineral deposition" attributed to the smallest PPi dose (Fleisch et al. 1966).

Beginning around 1970, the cultured chick femur results were revisited from several different approaches using a different analytical techniques and animal species.

Using radio-labeled calcium pre-administered to newborn mice during the growth phase, the effect of inhibitors of calcium loss could be quantified (Reynolds et al. 1972, see below). The laboratory research interests had also shifted to an intense focus on a new class of compounds whose "increased deposition" effect could be more easily discerned. By 1970, the group had tacitly adopted the concept that the increased "deposition" of calcium in living tissue was, in fact, an "antiresorptive" event allowing bone mass gain *relative* to the uninhibited resorbing controls. The effect was deemed to be cell-mediated.

---

[2] The authors thank the editors of the *American Journal of Physiology* for permission to reprint this historically important passage.

One model (Russell et al. 1970) was designed to measure the inhibition of the calcium increase effected by parathyroid hormone (PTH) given to (TPTX) rats. PTH was understood to mobilize calcium from the skeleton by cell mediated resorption; the object of this portion of the screen was to assign antiresorptive properties to substances which blocked the mobilization thus fulfilling the function termed "antiresorptive." The exact protocol is described in a somewhat piecemeal fashion but it proves decipherable. Unfortunately, the protocol was timed in such a manner as to fail to detect any effect of PPi. The animals were pretreated by subcutaneous injection for three days with a variety of potential inhibitors and *on the fourth day* were challenged with PTH. The rise in calcium with this challenge was not blocked by PPi. The diphosphonates were very effective, in this protocol, in blocking the PTH response. This effect was described as lasting "several days," most likely the interval before the test animals were discarded. The group had already expressed their opinion (Jung et al. 1970) that the effects of PPi would be short-lived.

The new diphosphonic acids of interest were structurally analogous to PPi. In the figures below, the acid forms are illustrated. At physiologic pH a portion of the hydroxyl groups would be ionized (the angled line in Figure 2 is the standard representation of the 2 bonds to phosphorus of the methylene group. Carbon-hydrogen bonds are by convention omitted). The bisphosphonic acids of medical interest have additional substituents replacing the hydrogens on the ($-CH_2-$) group.

The misconstruction of antiresorption of bone mineral (more mass retention than untreated controls) as increased stimulation of bone growth eventually was recognized by the Fleisch group. By 1972, this was received knowledge and another ex-vivo experimental protocol (mentioned above) was published (Reynolds et al. 1972) using newborn mice hemicalvaria, each corresponding half serving as its own control. This paper endorses the newfound cell-mediated antiresorption as the mechanism of the effect of the two bisphosphonates.

It is not the purpose of this review to document these bisphosphonate studies, although portions may be pertinent to the following chapter. However, these later papers contain some significant information pertinent to this PPi review.

There was increasing recognition in papers from this period that the effect of the bisphosphonate compounds now being tested was cell-mediated. The authors can find no initial announcement by the Fleisch research group accepting this mechanistic understanding converting the concept of (a) stimulation of crystallization (an essentially inorganic idea) to (b) inhibition of cell-mediated resorption, thus producing a similar gain in mineral mass over living controls since the controls had the same resorbing cellular effects in place. The effect was no longer considered to be directly on the crystalline structure of bone.

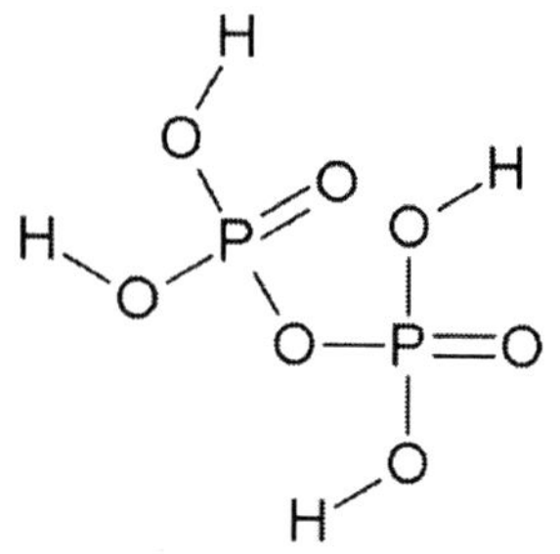

Figure 1. Pyrophosphoric Acid.

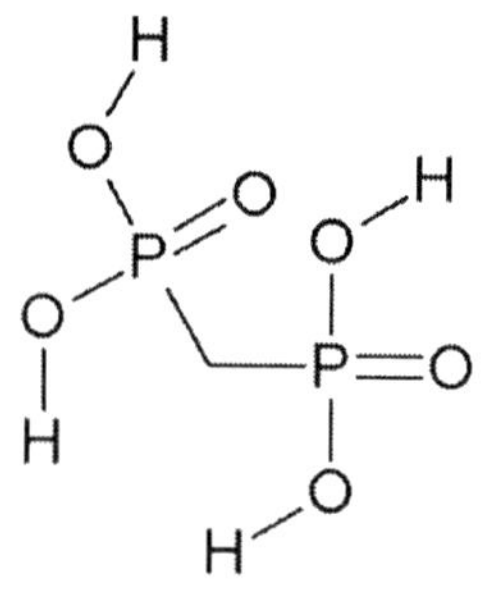

Figure 2. Methylene diphosphonic (bisphosphonic) acid.

Remarkably, it had been exactly a century since the astonishingly observant Swiss-born genius Albert Tölliker had discovered the osteoclast (von Tölliker 1873) and introduced the concept of cell-mediated resorption of bone. A direct effect on the deposition of crystalline bone had been the entire focus of the Fleisch work done with PPi, progressing as it had from inorganic preparations to ex-vivo studies to living explants and finally, in the case of the bisphosphonates, to pretreatment of living animal models. Testing of PPi parallel to this last stage was deemed to be likely unproductive since PPi was deemed to be unstable in the blood (Jung et al. 1970). This idea derived from a study which demonstrated the very rapid disappearance of labeled PPi from the serum into an unknown "second compartment." In dogs, 16-22% of intravenously administered labeled PPi was shown to be hydrolyzed to orthophosphate during the ~1 hour experimental interval and 4.4-11.3% of intact PPi was excreted through the kidneys. This averages to roughly 22% of the label accounted for, using the median values of the two ranges. The authors admitted that they could not account for the remaining ~78% of the PPi but opined that some of it "may be hydrolysed, although it is impossible to determine how much"!

In any case, the diphosphonate-focused experiment (Reynolds et al. 1972) referred to above consisted of injecting radio-labeled calcium into newborn mice, allowing a growth phase, and then isolating the living calvaria in an ex-vivo study with each hemicalvarium serving as a control the other side. Calcium loss into the medium was postulated to be due to resorption. In addition to the two diphosphonates tested, so was PPi. Even more interesting, is that the only dose chosen for the PPi was 4 μgms P/ml, a dose which showed only mineralization inhibition in the chick experiments (or at least the mineralization inhibition overcame the simultaneous antiresorptive effect). With the new understanding of the mechanism of action of the process leading to increased mineral mass, the 1 μgm P/ml dose of "inhibitor" likely would have revealed an effect essentially parallel to the diphosphonates. Such an assumption does not rest solely upon the proper interpretation of the chick results, but also on the findings of the antiresorptive power of PPi in two other unassociated laboratories, which will be described below.

However, there was one more effort from the Fleisch laboratory that will complete this portion of the review. It was postulated, perhaps rightly, that the "second compartment" into which injected PPi rapidly migrated had not been defined. The attempt to find this compartment involved incubation (Jung et al. 1973) of suspensions of PPi-free powdered hydroxyapatite, the form of calcium phosphate generally taken in bone, independently and competitively with PPi and several diphosphonates. PPi had the strongest binding. This led to the tacit conclusion that bone was the second compartment in the intravenously-injected dog experiments. This aqueous inorganic chemistry approach recapitulates the older series of studies of interaction on the physical chemical level alone.

This last paper (Jung et al. 1973) from the Swiss group in this review did contain an echo of the impression Fleisch had when commenting on the duality of the effect of PPi in the chick femur studies (Fleisch et al. 1966). Publications by Hajimi Orimo and Harold Rasmussen are referenced and the antiresorptive role of PPi is tacitly acknowledged. It remained for these two distant groups to confirm the antiresorptive activity of PPi.

THE EVIDENCE OF HAJIMI ORIMO THAT PYROPHOSPHORIC ACID HAS ANTIRESORPTIVE PROPERTIES

In order to demonstrate and confirm an antiresorptive character of PPi, protocols had to be designed that compensated for the rapid peripheral phosphatase-mediated hydrolysis of PPi possibly by presenting sufficient inhibitor to overcome the hydrolysis effect, bind to hydroxyapatite, survive local hydrolysis in the bone compartment, and exert the antiresorptive effect seen at very low levels in the earlier chick femur experiment of Fleisch. Alternately stated, the timing of the experiment had to take into account the transient nature of the antiresorptive window. The fact that these "requirements" were not present when the chick femur results of Herbert Fleisch actually detected an antiresorptive effect (albeit misunderstood) suggests that there was reason for an optimistic expectation for success if more attention was devoted to these details.

Hajimi Orimo approached the problem of defining the effect of PPi on bone metabolism from the point of view of endocrinology, thus his work was consistently in living systems resembling the last portion of the Fleisch studies before Fleisch committed entirely to the diphosphonate antiresorptive entities. The assumption that bone resorption was cell-mediated was understood in this group from the beginning of this work.

In the intact or TPTX rat, subcutaneous thyrocalcitonin (calcitonin) and PPi each produced (Orimo, Fujita, and Yoshikawa 1969) a decrease in serum calcium, but at the dose where the effect of calcitonin was maximal, PPi still reduced the calcium more. The greatest effect was detected at 30 minutes, but the effect dissipated over 5 hours. Obviously the effect of PPi by a parenteral route was limited, and measurement had to be timely to detect this augmentation. The ~24 hours allowed in the Fleisch 1970 experiment missed this effect. Orimo felt this experiment supported but did not conclusively prove that PPi extended the antiresorptive effect of calcitonin.

In another paper [Orimo, Fujita, and M. Yoshikawa. 1969a], Orimo described injection of pregnant mice with labeled ^{45}Ca to obtain labeled newborn calvaria. These were attached to coverslips by clotted chicken plasma and covered with a physiologic solution which was renewed and oxygenated every other day for a week, thus maintaining the cellular viability in the bone. Aliquots of the medium were counted at each exchange by scintillation. PTH increased the release of the label and this release was inhibited by PPi at 2-16 μM concentrations in the tube in a dose-dependent manner. Calcitonin inhibited the release to a somewhat greater extent. Combined PPi and calcitonin produced more inhibition than either of the agents alone, as had been the case in the intact rat experiments mentioned above. The very similar (Reynolds et al. 1972)) labeled calcium preparation reported in 1972 by Fleisch measured only the spontaneous release of labeled calcium. No PTH-stimulated release was studied. In that preparation, as noted above, the spontaneous release in the presence of 4 µgms P/mL of PPi (the only dose studied) was the same as that from live controls. In conclusion, Orimo's reported studies were consistent with PPi having antiresorptive properties.

EVIDENCE FROM THE LABORATORIES OF HOWARD RASMUSSEN

One of the present authors (AFD) in collaboration with the American physiologist Howard Rasmussen, completed a series of studies with results consistent with an antiresorptive role of PPi. A similar TPTX rat model was used with determination of the effect of PPi on the cell-mediated increase in calcium excretion effected by intravenous PTH. These carefully designed studies illustrate the importance of the timing of the steps. The intravenous route was chosen for the administration of the PPi. The reader will recall the variation of opinion of the researchers on the degree of stability of PPi. The subcutaneous route could well lead to a slow and measured introduction of the drug into the circulation, a circumstance that could increase the percentage of the drug that underwent hydrolysis. On the other hand, the very rapid uptake into bone noted in the dog intravenous studies (Jung et al. 1970) might well serve to improve the dose-effectiveness of PPi by limiting the portion initially undergoing to hydrolysis.

After testing (Rasmussen et al. 1970) simultaneous intravenous infusion of PTH + PPi, Rasmussen found (Delong, Feinblatt, and

Rasmussen 1971) that beginning the PPi infusion prior to the PTH challenge allowed the PPi suppression of calciuria to be better demonstrated. The control animals were treated with either PPi or PTH. In these controls, PTH caused an immediate rise in urinary orthophosphate but first invoked a slight drop in urinary calcium and then a progressive rise in urinary calcium and hydroxyproline. Urinary hydroxyproline was recognized (Anast et al. 1967; Kohler and Pechet 1967) at the time as a marker of bone resorption. The delay in calcium excretion was attributed to the direct effect of PTH increasing renal calcium retention, an effect now well established (Blaine, Conchol, and Levi 2015).

PPi (pH 7.4) was infused at 15 μmols/hr and after 4 hours of this pretreatment, PTH was simultaneously infused at 5 μgms/hr. The rise in urinary excretion of calcium and hydroxyproline observed in the PTH control was blocked. Continuation of the PTH infusion after stopping the PPi returned to the same rise in calcium excretion observed with the PTH-only, showing that as an antiresorptive, PPi was reversible. This met exactly the same standard applied by Fleisch to establish the antiresorptive activity of the diphosphonates, except that the unmetabolizable nature of the latter compounds precluded the reverse phase. The antiresorptive effect of the diphosphonates were described as lasting "a few days"!

In our (AFD) experiments, the rate of urine production was not affected by PPi. The inulin clearance was used to measure the GFR, which also was not significantly affected by PPi. This protocol compensated for the rapid peripheral hydrolysis of PPi (phosphatase-mediated) by presenting sufficient inhibitor to overcome the hydrolysis effect, bind to bone surfaces, survive local hydrolysis in the bone compartment, and exert the antiresorptive effect seen at very low levels in the earlier chick femur experiment of Fleisch. It is also possible, as suggested above, that the IV route allowed a more abrupt encounter with the bone and thus secured a higher percentage of the active antiresorptive agent in the compartment where it was active.

Conclusion

Pyrophosphoric acid has antiresorptive activity. Why is this issue important today? Pyrophosphoric acid has found no place in the medical armamentarium. Or has it, by proxy? Another question might be asked: why does it seem important that PPi has, or has no, antiresorptive properties? Pyrophosphoric acid is the acknowledged model upon which the widely exploited bisphosphonate drugs have been developed. Some of this series have been used medically specifically because they proved to inhibit mineralization as well as remodeling. Etidronate is an example of this dual nature. It was a fair assumption from the earliest explorations of the bisphosphonate properties that they would be expected to share the characteristics of their model, with the exception, of course, of metabolic degradation, since the bisphosphonates are metabolically inert and chemically inert except under extreme non-physiologic conditions. The misapprehension that one of these PPi characteristics was *not* antiresorption was not universal in the Fleisch group(s) but there was a tendency in that direction, and even acknowledged leads were not followed up. This void provided an attractive opportunity for Orimo and Rasmussen and their collaborators to correct the misapprehension.

There are several legacy industrial diseases arising putatively from excessive exposure to PPi. Discovering the circumstantial evidence disclosed by such an exposure might be aided if the unusual side effects of the bisphosphonate were discovered somewhere in the absence of these synthetic products. The comprehension of such a relationship depends on a precise knowledge of the chemistry of elemental phosphorus, which is certainly not a common part of the medical or life sciences curriculum. Neither is the teaching of the details of bone metabolism found in chemistry curricula. This elusive interface will provide a subject for the following chapter.

A small portion of the above chapter has been abstracted from our previously published paper "An Evaluative History of Bisphosphonate Drugs: Dual Physiologic Effects of Pyrophosphate as Inspiration for a Novel Pharmaceutical Class. *Journal of Osteoporosis*, Volume 2016, Article ID 1426279, pp 1-7. doi:10.1155/2016/1426279.

REFERENCES

Anast, C., C. D. Arnaud, H. Rasmussen, and A. Tenenhouse. 1967. "Thyrocalcitonin and the response to parathyroid hormone." *The Journal of Clinical Investigation* 46: 57-64.

Blaine, J., M. Chonchol, and M. Levi. 2015. "Renal control of calcium, phosphate, and magnesium homeostasis." *Clinical Journal of the American Society of Nephrology* 10: 1257-1272.

Delong, Allyn, J. Feinblatt, and H. Rasmussen, 1971. "The effect of pyrophosphate infusion on the response of the thyroparathyroidectomised rat to parathyroid hormone and adenosine-3', 5'-cyclic monophosphate." *Calcified Tissue Research* 8, 87-95.

Fleisch, Herbert and W. F. Neuman. 1961. "Mechanisms of calcification: role of collagen, polyphosphates, and phosphatase." *American Journal of Physiology* 200: 1296-1300.

Fleisch, Herbert and Sylvia Bisaz. 1962. "Isolation from urine of pyrophosphate, a calcification inhibitor." *American Journal of Physiology*, 203: 671-675.

Fleisch, Herbert, Fritz Straumann, Robert Schenk, Sylvia Biaz, and Martin Allgöwer. 1966. "Effect of condensed phosphates on calcification of chick embryo femurs in tissue culture." *American Journal of Physiology* 211: 821-825.

Fleisch, H., R. G. G. Russell, and F. Straumann. 1966a. "Effect of Pyrophosphate on Hydroxyapatite and its Implications in Calcium Homeostasis." *Nature* 212: 901-903.

Harmey, Dympna, Lovisa Hessle, Sonoko Narisawa, Kristen A. Johnson, Robert Terkeltaub and José Luis Millán. 2004. "Concerted Regulation of Inorganic Pyrophosphate and Osteopontin by Akp2, Enpp1, and Ank: An Integrated Model of the Pathogenesis of Mineralization Disorders." *The American Journal of Pathology* 164: 1199-120.

Jung, A., R. G. G. Russell, S. Bisaz, D. B. Morgan, and H. Fleisch. 1970. "Fate of intravenously injected pyrophosphate-32P in dogs." *American Journal of Physiology* 218: 1757-1764.

Jung A., S. Bisaz, and H. Fleisch. 1973. "The binding of pyrophosphate and two diphosphonates by hydroxyapatite crystals." *Calcified Tissue Research* 11: 269-280.

Kohler H. F., and M. M. Pechet. 1967 "The inhibition of bone resorption by thyrocalcitonin." *Journal of Clinical Investigation* 45: 1033.

Neuman, William F. 1950. In: *Metabolic Interrelations* Josiah Macy Jr. Foundation 187.

Orimo, H., T. Fujita, and M. Yoshikawa. 1969. "Pyrophosphate enhancement of hypocalcemic effect of thyrocalcitonin in rats." *Endocrinologia Japonica* 16: 309-313.

Orimo H., T. Fujita, and M. Yoshikawa. 1969a. "Pyrophosphate enhancement of thyrocalcitonin effect in tissue culture." *Endocrinologia Japonica* 16: 415-421.

Orriss, I. R., T. R. Arnett, and R. G. G. Russell. 2016. "Pyrophosphate: a key inhibitor of mineralisation." *Current Opinion in Pharmacology* 28: 57-68.

Rasmussen, H., J. Feinblatt, N. Nagata, Allyn DeLong. 1970. "Regulation of bone cell function." In: *Osteoporosis* (Barzel, U. S., ed.) 187-198. New York-London: Grune & Stratton.

Reynolds, J. J., C. Minkin, D. B. Morgan, D. Spycher, and H. Fleisch. 1972. "The Effect of Two Diphosphonates on the Resorption of Mouse Calvaria *in vitro*." *Calcified Tissue Research* 10:302-313.

Russell, R. G. G., R. C. Mühlbauer, S. Bisaz, D. A. Williams, and H. Fleisch. 1970. "The influence of pyrophosphate, condensed phosphates, phosphonates and other phosphate compounds on the dissolution of hydroxyapatite *in vitro* and on bone resorption induced by parathyroid hormone in tissue culture and in thyroparathyroidectomised rats." *Calcified Tissue Research*, 6: 183-196.

Terkletaub, Robert A. 2001. "Inorganic pyrophosphate generation and disposition in pathophysiology." *American Journal of Physiology - Cell Physiology* 281: C1-C11.

Von Tölliker, R A. 1873. *Die normale Resorption des Knochengewebes und ihre Bedeutung fur die entstehung der typischen knochenformen* [*The normal resorption of the bone tissue and its importance for the development of the typical bone shapes*]. F. C. W. Vogel (Publisher), Leipzig.

In: A Closer Look at Pyrophosphates
Editors: Ritesh L. Kohale et al.
ISBN: 978-1-53617-730-5
© 2020 Nova Science Publishers, Inc.

Chapter 5

THE ANALOGOUS CAUSATION OF PHOSSY JAW, OSTEONECROSIS OF THE JAWS, INDUCED FEMUR FRAGILITY, AND ATYPICAL FEMUR FRACTURES

William Banks Hinshaw[1,*] and Louis DuBose Quin[2]
[1]Department of Surgery, Harris Regional Hospital,
Sylva, N. C., US
[2]James B. Duke Professor of Chemistry Emeritus,
Duke University Department of Chemistry, Durham, N. C., US

ABSTRACT

It has been generally acknowledged that the two sets of major side effects "associated" with bisphosphonate treatment, recognized since 2003, closely resemble two industrial maladies suffered by workers in the phosphorus match industry in the latter part of the 19th century.

---

[*] Corresponding Author's Email: williambh@frontier.com.

In the most striking of the legacy cases, severe disruption of the integrity of part of alveolar bones proper of the maxilla and/or mandible resulted in loss of teeth and skeletal jaw(s) and often was followed by superinfection leading to permanent disability and sometimes death. Because of its specificity to the phosphorus industries, this became known as "phossy jaw." Recognition of the similarity of phossy jaw to a virtually identical disruption associated with modern bisphosphonate therapy led to designating this side effect by several variations of the term "osteonecrosis of the jaws."

The more subtle of the legacy major side effects was an induced fragility of the long bones, chiefly the femur. This problem received less attention at the time, but was recorded and discussed by specialists concentrating on the phossy jaw problem chiefly because it was noted only in workers in the same phosphorus industries and differed from common fractures by the trivial force required to cause the breaks. The recent cases, however, have received much greater attention because of the severe disruption of the long-term quality of life in affected individuals due to total or threatened collapse of one or both femurs, ordinarily one of the strongest bones in the body.

Recognition of the similarities of these two sets of effects, sometimes occurring in a single individual, did not lead to the rationalization of any connection between the legacy cases and their modern equivalents. In the legacy situation the tendency was to blame the very poisonous elemental phosphorus which was utilized in all the processes where the side effects arose, even though somewhat more scientifically sophisticated minds recognized there were fundamental differences between elemental poisoning and the side effects described above. The modern cases may be associated with a molecule containing a phosphonic acid moiety, but all similarity between etiologies appears to end there. Elemental phosphorus plays no part in the production or the very restricted metabolism of these molecules.

In this review chapter, we discuss the conclusive modern analytical chemistry evidence that the true operating agent(s) in the chronic toxic exposures suffered by the legacy case workers is the same molecule (and precursor molecules) upon which the bisphosphonate drugs were modeled. Furthermore, the earliest studies leading the research teams to consider the bisphosphonates as potential medication demonstrated that this molecule and its precursors shared the antiresorptive effect on cell-mediated bone resorption that was discovered in the bisphosphonic acids. Three major research laboratories were involved in the studies of this molecule, which is, of course, pyrophosphoric acid. In the first and somewhat earlier of these groups, the antiresorptive character was discovered, commented upon, and then subsequently ignored, and, much later on, even denied. The remaining two totally independent laboratories were able to establish the antiresorptive effect through proper management of the subtle techniques required.

The search for animal models for the modern set of side-effects has been long and difficult but ultimately successful. Reproduction of the side effects in parallel experiments should be relatively easy in the case of chronic parenteral administration of pyrophosphoric acid. We show that there is evidence that a pulmonary route of administration is unnecessary. There are a number of physical parameters needing to be defined and prejudices to be overcome in order to understand the fate of pyrophosphate administered exogenously to living models. We feel confident that this demonstration can be successful.

INTRODUCTION

Osteonecrosis of the Jaws and Phossy Jaw

When it slowly became obvious, initially due to the herculean efforts of a single oral surgeon (Marx 2003), that the treatment of several metastatic malignancies (as well as postmenopausal osteoporosis) with the bisphosphonate drugs pamidronate and zoledronate was associated with a painful and debilitating dental condition that became known as osteonecrosis of the jaw (ONJ), it was not long before sporadic observations (Purcell and Boyd 2005) began to appear in the medical literature that the ONJ characteristics recalled an industrial disease, occurring most frequently in 19th century white phosphorus match factories workers, known as "phossy jaw" (Pollock, Rubin, and Brown 2015). This had been a devastating and often-fatal disruption of the alveolar bones, associated with loss of teeth, exposed dead bone, and, especially in the 19th century pre-antibiotic era, overwhelming infection. The later paper referenced immediately above is a rather florid exposition of the resemblance between the legacy "phossy Jaw" and the modern ONJ. We have included it because of its historical accuracy. These authors' opinions about etiology are their own. The reader is encouraged to consult this source.

Atypical Femur Fractures and Induced Femur Fragility

In the same year as the original ONJ publication, a poster presentation (Odvina et al. 2003), at an annual conference of the American Society of Bone and Mineral Research, disclosed an apparently similar relationship between induced fragility of certain specific bones in the human skeleton and the same class of bisphosphonate drugs. This did not garner the immediate attention of the former article (Marx 2003) because the publication of the extended material in the medical literature was delayed. In addition to that, while an equivalent 19th century match industry industrial malady existed for these fractures as well as for the jaw problem, it had not attracted the intense attention of the grisly jaw disease. Phossy jaw was a unique and never before encountered lesion, whereas broken long bones were commonplace. The unusual circumstances of occurrence of these fractures helped to draw the attention of the medical profession to them just as the more specific characteristics of the contemporary fractures place them in a class by themselves. Dr. Brocoorens in Grammont, Belgium, for example, reported (Oliver 1902) that in a small town with 1100 people employed in the match industry, 30 cases of fractures of the long bone caused by muscular effort alone had occurred over twenty-five years to individuals who had been diagnosed and recovered from phossy jaw. Dr. Dearden, from whom we shall hear more below, reported (Dearden 1899) two similar cases in match industry workers which he termed "Fragilitas ossium," an appropriate term which nonetheless did not persist and has come to imply a possibly entirely different syndrome in modern times. Dearden's cases were bilateral and "broken in a ridiculously simple fashion."

Two Pairs of Unusual and Easily Recognized Lesions, Related by Circumstance, Separated by More Than a Century

What elusive element connects these events? How can two industrial diseases satisfactorily eliminated a century ago (by finding an alternative to white phosphorus in the manufacturing process) reappear still yoked together by an element in common: phosphorus match factory work in the legacy cases and bisphosphonate therapy in the modern ones? There have been widely varied explanations ranging from well-reasoned pathways of causality starting from indeterminate or unproved exposures as well as the most naïve but nonetheless smugly confident concoctions claiming to settle the matter once and for all. However up to 2007, there had been no convincing uniform theory of causality, starting from existing well-established results of what exposure to burning white phosphorus entailed, that reconciled the two 19th century lesions with their two 20th century parallels. At that time, the two present authors consulting together realized that there was an extant body of analytical chemistry studies which, combined with the eye-witness descriptions of what occurred regularly in the manufacturing of the white phosphorus matches, and, combined with what we know about the effects of the known combustion products on bone, provided entrée into a comprehensive theory (Hinshaw 2007) of the quadruple causality. By far the most important factor was the identification of substance(s) to which the victims had been exposed. This is not very difficult to determine, as the ultimate reaction products with air of elemental phosphorus have been fully determined by multiple researchers. Early guesses at this fundamental cause ranged from the volatized element itself through its known primary oxidation products. Most of speculation took place in an era where the constitution of burning phosphorus smoke had not been defined. Once this was understood, the relationship could be understood. And if, perchance, this product(s) had already been known to have a strong pharmacodynamic impact on bone, then a solution to the ancient controversy was at hand.

Social Responses to the Legacy Problems

The problem of phossy jaw affected the least influential class of the working population, but as it became recognized, the private sector and governmental authorities both began to press for measures of prevention. Judging by the length and number of the reports, the social response was vigorous. This produced detailed descriptions of the working conditions of phosphorus match manufacture recorded in official governmental investigation reports (Thorpe, Oliver, and Cunningham, 1898) which describe frequent flare-ups of spontaneous ignition especially in the "dipping" rooms where arrays of the wooden stems were semi-automatically dipped into the pastes containing wax, white phosphorus, chlorate, and other material. Even more common in the "boxing rooms," where the dried double-head matches were sliced and packed by hand into metal containers, small fires were commonplace, usually beaten out by hand by the young woman employed to do the boxing (Oliver 1902). While it may have been natural to suspect the elemental phosphorus itself as the culprit in this condition, since it spontaneously fumes heavily when exposed to air, the evidence that has come down to us suggests otherwise. Physical contact with the element induces severe dermal burns and ingestion of even a small amount can prove fatal (Emsley 2000). Workers in the industry were encouraged to scrupulously scrub their hands and arms after work, and especially before meals. The victims of phossy jaw frequently succumbed to systemic infection following the disruption of their oral mucus membranes in a pre-antibiotic age, but those who did recover, especially after the value of topical antiseptics became accepted, while permanently deformed, did not appear to suffer the specific consequences of phosphorus toxicity, such as the complete collapse of liver function. The smoke per se, however, was suspected to be the causative agent, especially when the introduction of strong air flow away from the workers was observed to reduce the incidence of the problems (Oliver 1902).

CHEMISTRY OF BURNING WHITE PHOSPHORUS

An understanding of the details of the chemistry of inorganic pyrophosphoric acid (PPi) presupposes that the reader is familiar with the preceding chapter (Hinshaw and DeLong 2020).

Chemistry is not a speculative science. The theoretical prediction of the course of chemical reactions is of course subject to error and failure, but once a set of reaction conditions and the products thereof has been definitively determined, there is no room for speculation about subsequent exemplifications. The same is true of prediction of the course of unobserved reactions that have no chemical precedent or logic. Chemistry is defined by sets of generally accepted rules developed over many years about the limitations imposed by natural constraints on how reactions may proceed, and the mere stoichiometric balance of reaction elements and products is no guide to whether such a process is probable or to be expected. There have been several conjectures (Marx 2008) regarding the causation of the legacy diseases being a bisphosphonate formed in the bodies of the match industry victims. Postulated paths to these agents are not consistent with the contemporary science of biochemistry or chemistry! There is no evidence that the carbon-phosphorus bone can be biosynthesized in mammalian systems, although 2-aminoethylphosphonic acid, a compound richly present in some edible Crustacea, has been isolated from some human tissues in tiny amounts, thought to be the result of ingestion (Kukher 2000).

The chemistry of burning white phosphorus in atmospheric conditions is now understood to terminate in orthophosphoric acid, but the rates of the various steps of the process vary greatly. Some are virtually instantaneous and others have a slow rate of proceeding under normal conditions of temperature and water content of the air. Elemental white phosphorus is a waxy solid at room temperature but with a very low melting point. Nonetheless, its vapor pressure at room temperature is only 3% of that of ice at the freezing point. The white form of elemental phosphorus exists as a highly strained tetramer (Corbridge 1978). It is so reactive with oxygen that a piece exposed to air unprotected rapidly bursts into flame. The survival of a molecule of P_4 escaping the surface of the solid is generally estimated to be so subject to immediate oxidation that the element itself does not exist very far away from the mass, whether it is simply subliming or has begun to burn. This means, very pertinent to our analysis, that except for explosions scattering the elemental mass, which were not reported in the legacy investigations, that workers were exposed in the smoke to essentially no elemental white phosphorus. One analysis of the air around a P_4 process found (King and Peterson 1978) concentration of at most 0.01 mg/ cubic meter. The lethal dose of 50 mg is thus 5000 times that high.

Thus elemental phosphorus is not found mixed with the products of its oxidation in the analyses described below. In the presence of ample air the primary oxidation product is predominantly P_4O_{10} (van Wazer 1966). Again neither P_4O_{10} nor the other known oxide is found among the hydrolysis products which form the smoke. When the definitive experiment was published by the US Army Chemical Corps in 1985 (Spanggord et al. 1985), only the penultimate hydrolysis products could be detected. In timed analyses, Spanggord examined the content of the smoke from burning white phosphorus using gas-liquid chromatography at intervals immediately after firing, and at 4 hours, 24 hours and 96 hours. The initial results revealed that the mixture consisted of roughly 25% of the terminal fully hydrolyzed inert orthophosphate, 25% of pyrophosphoric acid, and 50% of a mixture of the linear phosphorus polyacids P_3 through P_{15}. As the smoke ages in the presence of atmospheric humidity, the content of the polyphosphoric acids progresses through the smallest polyacids, PPi, to the ultimate orthophosphate. This process is illustrated below. Step 1: White phosphorus very rapidly reacts with oxygen in the air to generate the oxide. Step 2: The oxide reacts rapidly with water (hydrolysis) in the air. The reaction goes to completion and includes reaction between some of the intermediates to produce phosphoric acids with between one and 15 phosphorus atoms. Tripolyphosphoric acid is illustrated (P3 on the charts below). Step 4: All hydrolysis steps pass through the next to the last hydrolysis product which is pyrophosphoric acid (P2 in the charts below). The rate of the hydrolytic steps is much slower than the prior stages of this sequence. Step 5: The hydrolysis of pyrophosphoric acid produces the hydrolytically-stable orthophosphoric acid.

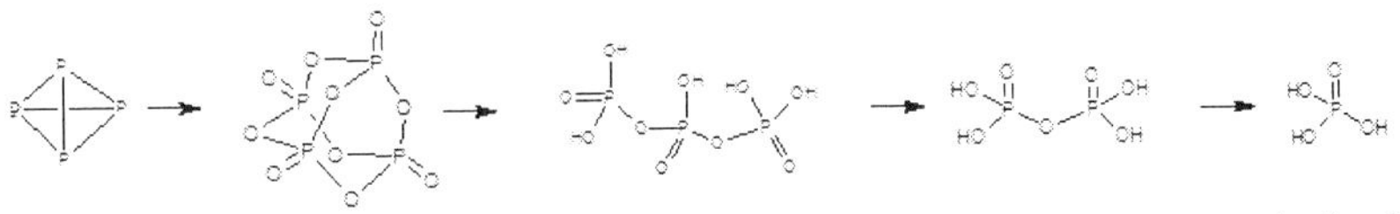

| White Phosphorus Acid | P_4O_{10} P(V)Oxide | Tripolyphosphoric Acid | Pyrophosphoric Acid | Orthophosphoric Acid |

Figure 1. The production of hydrolytically-stable orthophosphoric acid from hydrolysis of pyrophosphoric acid.

These Army studies, after decades of speculation, defined the smoke that was being inhaled by the workers in the most exposed job positions as mentioned above. It is perhaps not a coincidence that the active ingredients of the smoke (which excludes the nutritional orthophosphate) are the same as those tested by Fleisch ex-vivo regarding effects on growing bone (Fleisch et al. 1966): pyrophosphoric acid and a mixture of the lower polyacids. Both of these entities demonstrated antiresorptive effects in the experimental tissues. As described previously (Hinshaw and DeLong 2019), an appreciation of this effect was delayed.

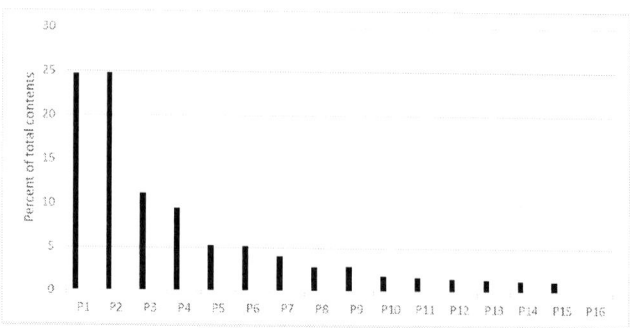

Chart 1. Smoke Initial Content.

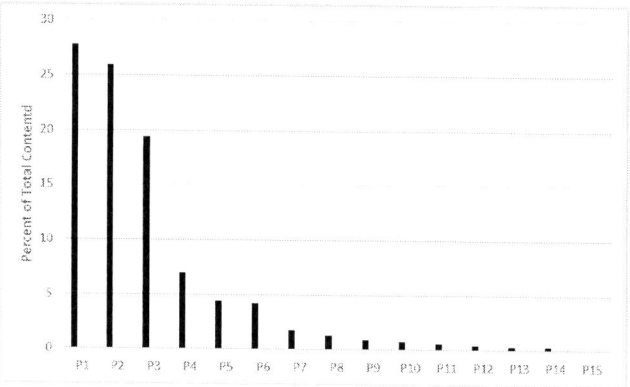

Chart 2. Smoke Content after 4 hours.

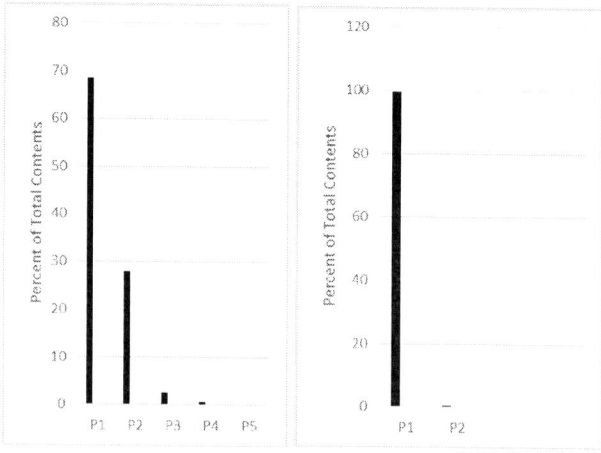

Chart 3, Chart 4. Smoke Content after 24 hours Smoke Content after 96 hours.

From the above data, it may be seen that the PPi (P2 in the charts) percentage content of the smoke does not change for the first 24 hours, as hydrolysis of the longer polyacids replenish the supply of PPi even as the PPi is itself being hydrolyzed to orthophosphate. In the legacy analyses it was deduced (Oliver 1902) that exposure to the fresher the smoke was associated with a higher incidence of phossy jaw, but this may have been a volume issue as the smoke would likely be disbursed long before its PPi content decreased.

PPi and the polyacids are soluble in water, even more so at physiologic pH. Absorption via the pulmonary route seems very probable. There are some data available to shed light on this question. PPi and ^{99m}Tc-Sn-PPi have been shown (Bisaz, Jung, and Fleisch 1978) to behave very similarly with respect to uptake in bone after intravenous injection. Similarly, ^{99m}Tc-Sn-PPi has been used (Isitman et al. 1988) for pulmonary scanning in human volunteers, by inhalation. About 70% of the retained label has disappeared from the lungs at 6 hours but only 3.3% of the activity was detected in the blood at 30 minutes and 3.9% at one hour. This latter apparent discrepancy between the label entering the blood from the lungs and the label measurable in the blood is quite consistent with Fleisch's observation (Jung et al. 1970) that labeled PPi injected in dogs is very rapidly taken up into a second compartment, exiting the blood. All these data are consistent and supportive of systemic absorption of PPi by the pulmonary route in the legacy cases.

EVIDENCE FOR BINDING OF PYROPHOSPHATE TO BONE *IN VIVO* AND POTENTIAL CONSEQUENCES

William Dearden, practicing public medicine in Manchester in the last years of the 19th century was convinced that the match industry lesions he had seen and reported were not due to elemental phosphorus. Efforts in several laboratories to implant small quantities of the element in the jaws of dogs had not produced similar effects. Dearden argued, without any experimental basis, that the smoke from burning phosphorus likely contained the oxides, perhaps even a surplus of the P(III) oxide P_4O_6 (tetraphosphorus hexoxide) based on putative (and unproved) low levels of oxygen available in the factories. Dearden assumed, again without real foundation, that this compound would be more reactive, and thus more dangerous, than the P(V)oxide P_5O_{10} (pentaphosphorus decaoxide), which encouraged him in his choice. As noted above, it was not until 1985 that the experimental evidence eliminated these moieties as components of the smoke.

One of Dearden's patients who had experienced two atraumatic leg fractures had the misfortune to lose the distal phalanx of a finger. Dearden was able to obtain a chemical analysis of the bone and to compare this to analysis of a finger of a man of the same age who was not employed in the phosphorus industry, this control specimen having been obtained at autopsy. Dearden reports a slight increase in the phosphorus-to-calcium ratio in his afflicted patient compared to the autopsy specimen and the contemporary accepted values in the medical literature. This simple experiment confirmed Dearden in his belief that the exposure over the long term to the phosphorus smoke had induced an intrinsic change in the character of the bone which systemically contributed to the lesions.

At the time, the German scientist and physician von Steubenrauch in Munich expressed (von Steubenrauch 1899) his conviction that neither implantation of the oxides nor the element nor a variety of other phosphorus compounds nor long residence of the experimental animals in the smoky boxing rooms could produce a model for the phossy jaw and fragile femur syndromes. Working barely 5 years after the discovery of X-rays by Roentgen, he performed fluoroscopic (von Steubenrauch 1900) analyses on three long-time workers in the match industry. We obtained photocopies of the original article, entitled "About the changes of growing bone caused by the influence of phosphorus," from a library in Munich, but von Steubenrauch's reports contain no reproductions of images. He describes dense bands in the epiphyses of the tibia and radius of one subject and in the fibula and radius of two others which are reminiscent of those seen in experimental animals (Schenk et al. 1973). Dark bands near the epiphyses of the long bones, which were poorly calcified areas of resorption, were illustrated in animals treated with bisphosphonates.

The Fleisch group (Reynolds et al. 1972) also described similar lesions in mice chronically treated with high-dose bisphosphonates (BP). The results were attributed to severe inhibition of bone resorption resulting in failure of development and early death. Fleisch had predicted that the chronically BP-treated mice would resemble a hereditary mouse strain known as grey-lethal osteopetrosis which had a very short life span due to inherited failure of bone resorption. He was able to demonstrate metabolically and radiologically that this was the case in the BP-treated mice as well. Osteopetrosis is a hereditary syndrome with several recognized variants which is characterized by defective bone resorption leading to a very dense skeleton susceptible paradoxically to frequent fractures caused by the brittleness of the bone, as well as jaw complications resembling ONJ (Lam et al. 2007, Yates et al. 2010).

An inadvertently induced illustration of this BP effect in a human being has been followed long-term by Michael Whyte. A boy who was treated with high dose pamidronate (a bisphosphonate) developed characteristics of genetic osteopetrosis, including spontaneous bone fragility (Whyte 2003).

DEVELOPMENT OF BISPHOSPHONATE DRUGS BASED ON PYROPHOSPHATE

The earliest developments regarding the pharmacology of PPi were concentrated in the research facilities of the Swiss biochemist Herbert Fleisch. We have reviewed his contributions in the previous chapter

(Hinshaw and DeLong 2020). Fleisch soon abandoned the pursuit of the biological potential of exogenous PPi after he was introduced to the analogous bisphosphonate compounds by Marion D. Francis from the Proctor & Gamble Company (Fleisch et al. 1968). Earlier, Fleisch had decided, based on weak evidence (Jung et al. 1970), that exogenously administered PPi could not survive hydrolysis long enough to produce measurable effects in vivo. The new bisphosphonates proved to be resistant to hydrolysis and shared a characteristic that he had sought in PPi: blocking the cell-mediated resorption of bone. We have reviewed the efforts of Fleisch and other scientists elsewhere (Hinshaw 2007; Hinshaw and Quin 2013; Hinshaw and Quin 2015; Hinshaw and DeLong 2016). There has been some effort (Orriss, Arnett, and Russell 2016) to disparage the idea that PPi has antiresorptive properties. These efforts are misinformed, as we have shown in the preceding chapter. This issue may relate to the fact, now established, that the bisphosphonate antiresorptives, universally acknowledged even in the labels to be pharmacologic analogs of PPi, can and do function in vivo as antiresorptives in a manner parallel to PPi, combining antiresorptive and mineralization inhibition properties. Etidronate is an example of an approved bisphosphonate drug which can be used medically because of its predominant anti-mineralization character (Fleisch 1998). The antiresorptive character of PPi completes the analogy with the BPs and leaves open the possible predictability of ONJ and atypical femur fractures (AFF) as side effects of the modern drugs based on the legacy exposure to PPi.

The expectation, that the bisphosphonate structure might occupy enzymatic or crystal lattice sites competitively or non-competitively was based on the molecular similarity to the pyrophosphoric acid, had been intensively investigated by the Fleisch group before 1970. Figure 2 below illustrates the similarity between the simplest bisphosphonic acid (methylenediphosphonic acid or MDP) on the left and PPi on the right.

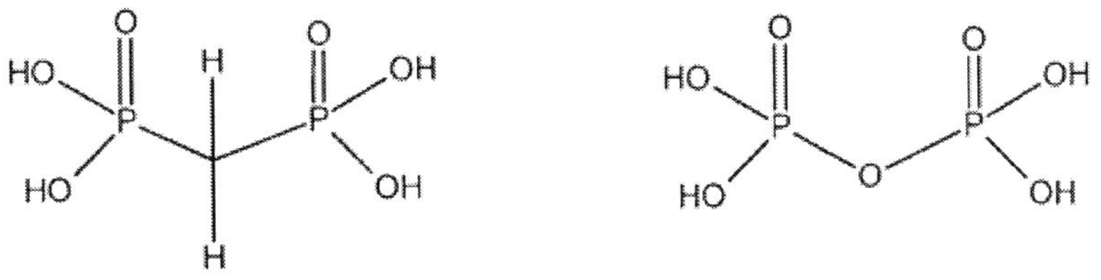

Methylenebisphosphonic (diphosphonic) Acid **Pyrophosphoric Acid**

Figure 2. The similarity between the simplest bisphosphonic acid (methylenediphosphonic acid or MDP) on the left and PPi on the right.

The Modern Response to Osteonecrosis of the Jaws

The publication of the original cases of ONJ (Marx 2003) was followed by hundreds of cases reported to the medical and dental literature. The first peer-reviewed article appeared the next year and described cases occurring between 2000-2003. The interest and concern generated prompted the American Society of Bone and Mineral Research to convene a task-force of recognized experts to address the problem. The report (Khosla et al. 2007) of this committee did not acknowledge the bisphosphonates as the sole etiology but did take note of a 1-10% incidence of ONJ in oncology patients receiving the medicine intravenously. The incidence was noted to be very much lower among users of the drugs by the oral route. The question of causation did not seem to be very much debated, due to the uniqueness of the lesion, but it was not acknowledged in this report. The dental community rapidly adopted extreme caution in dealing with patients who had taken these drugs intravenously, although many of the avoidance procedures proposed (Ruggerio et al. 2007) manifested a degree of ignorance of BP persistence and mechanism of action. Dental recommendations (Edwards et al. 2007) for persons having taken oral BPs have been much less restrictive.

THE MODERN RESPONSE TO THE ATYPICAL FEMUR FRACTURES

The denial of causality by the BPs for the AFFs has been much more pronounced. Perhaps this is to be expected, since the lesion, despite its unique anatomical characteristics, inevitably overlaps with femur fractures of traumatic origin. This same issue would certainly have limited the recognition of the legacy fractures, which might well have gone unnoticed had the association with the unique phossy jaw not served to point to a common cause. The AFF lesion has again been formally addressed by the American Society of Bone and Mineral Research, this time by two task force reports with overlapping authorship. The first report (Shane et al. 2010) defined the fractures very tightly, so tightly in fact that the second report (Shane et al. 2013) modified the original definition. Neither of these reports acknowledged the bisphosphonate drugs to be the cause of the fractures, but the second report did include the statement that "Although the task force still holds the opinion that a causal relationship between BPs and AFFs has not been established, evidence for an association has continued to accumulate in the 2 years since the first report was published and is quite robust. Moreover, the fairly consistent magnitude of the association between BPs and AFFs is unlikely to be accounted for by unknown or unmeasured confounders." Part of this difficulty doubtless stems from the problem of making any acceptable statements about causality, a conundrum which tried the abilities of such great philosophers as David Hume. The scientists and physicians who produced these documents were doubtless divided on the issue of causation and chose not to confront it. The problem has been addressed in a rational manner by Sir Austin Bradford Hill (Bradford Hill 1965) and his method has been accepted by the U. S. Food and Drug Administration (U.S. Department of Health and Human Services Food and Drug Administration. 2005) for resolving similar conundrums. This methodology was not applied by the Task Force groups.

CAN HYPOPHOSPHATASIA BE EXPECTED TO CAUSE BONE EFFECTS ANALOGOUS TO ONJ AND AFF?

Hypophosphatasia is an inherited metabolic disorder which was first generally described in 1948 as an independent entity (Rathbun 1948) and well characterized over the following decade (Fraser 1957). This disease has variable degrees of penetration from being lethal in childhood to being barely recognizable later in life.

In 1971, Fleisch and Russell, having devised an elaborate analytic technique for blood PPi, undertook an investigation (Russell et al. 1971) of the PPi levels in individuals diagnosed with osteogenesis imperfecta, osteopetrosis, "acute" osteoporosis, primary hyperparathyroidism, and hypophosphatasia. Only in the last mentioned group were PPi levels found to be consistently high compared to their control group. This was, one supposes, to be expected since the disease had been long known as stemming from reduced activity of alkaline phosphatase, but perhaps it did add to the conviction that PPi was a likely substrate for the enzyme.

Fleisch and Russell never commented on the implications of treatment with PPi analogs resistant to hydrolysis (bisphosphonates) possibly engendering hypophosphatasia-like syndromes, in that the antiresorptive replacing PPi would be analogously resistant to enzymatic hydrolysis as is the PPi when the enzyme is the inadequate part of the required process. Hypophosphatasia is characterized by susceptibility to fracture, a fact that they attributed to the mineralization-inhibition part of PPi functions. Recognition of the similarities between hypophosphatasia and the bisphosphonate side-effects was only achieved in more recent times. In 2009, Michael Whyte pointed out (Whyte 2009) that the growing reportage of an association between bisphosphonate therapy and what was becoming known as the atypical femoral fractures recalled that "for at least 30 years" similar fractures had been known as "the hallmark of the adult form of hypophosphatasia." In his short and very incisive "clinical vignette," Whyte attributes the fractures in adult hypophosphatasia to osteomalacia or bone formation failure, associated with the known PPi mineralization inhibition, to which the patients have, of course, been long exposed. He then reasons that unmasking of hypophosphatasia by the bisphosphonate treatment is "probably not" the cause of the BP-associated AFFs and that the occurrence of these fractures are likely to result from the pharmacologic action of the BPs which suppresses bone turnover. There can be no doubt, based on the material presented in the previous chapter (Hinshaw and DeLong 2019) that the pharmacologic action of the PPi to which these patients are subject includes antiresorptive activity which similarly suppresses bone turnover. Data clearly suggesting this possibility was available since the early 1970s.

Several years later, Whyte did indeed publish a detailed case report (Sutton et al. 2012) of a woman who was given two bisphosphonates based on a diagnosis of low bone mass, who developed several metatarsal fractures (see also Hinshaw and Schneider 2018) in the first several months after the start of therapy and bilateral subtrochanteric atypical femoral fractures after 4 years. Afterwards, genetic analysis of the patient revealed a low level of penetration of the adult form of hypophosphatasia.

While these data were interpreted in the light of osteomalacia due to PPi exposure complicated by the antiresorptive effects of the BPs, it is equally possible to replace such a "supplementary" hypothesis with an "additive" one. The fact is that in normal young adults working in the phosphorus match industry, significant repetitive dosing with exogenous PPi led to these same sort of fractures, with no documented associated osteomalacia.

MECHANISTICALLY-COMPETENT SCIENTISTS ACKNOWLEDGE THE RISKS OF BISPHOSPHONATE DRUGS

Medications deemed useful for therapy are often designed to affect certain known systems in certain ways to yield a desirable effect. Even if that goal is achieved, the assumed design (often called the "mechanism of action") may later prove to be wrong. Concurrently with the growth of knowledge of bone remodeling and self-repair, the loss of bone mineral proved an attractive target for therapy because of the relative simplicity of measurement. This impression accelerated, under the marketing influence (Spiegel 2009) of the major producers of the bisphosphonate drugs to the medical establishment, with the growing emphasis on bone mineral density mensuration. Herbert Fleisch proved to be the leading advocate of the bisphosphonate drugs as the solution to the search for a long-lasting antiresorptive agent which would increase bone mineral density and possibly the mass of bones deemed to have insufficient mineral. But the limitations of a mechanism of action based on improving the mineral content of bone at the expense of restriction of the remodeling repair process became apparent to him years later. In a definitive review of the state of the knowledge of the bisphosphonates, in a section regarding the physical (mechanical) properties of bone, Fleisch warned (Fleisch 1998b) "This issue is important since long-lasting, strong inhibition of bone resorption can lead to increased bone fragility and, therefore to fractures caused by the inability to replace old bone by young bone and to repair microcracks."

A. Michael Parfitt made contributions to the understanding of bone metabolism that place him among the most perceptive of the scientists and physicians in this field. It is significant that four of the most respected scientists (Manolagas et al. 2015) in this field attributed, in a long and detailed encomium, the following to Parfitt. "Non-mass factors together contribute more to bone fragility than does loss of bone mass, which undermines the continued use of the WHO criteria for the so-called "diagnosis" of osteoporosis." The "WHO criteria" refers to the use of the arbitrary bone mineral density limits defining who can be said to have osteoporosis. The year after the FDA approval of the first bisphosphonate, he warned (Parfitt 1996) that the present state of knowledge "did not eliminate the possibility" that after 10-20 years there could be "an epidemic of hip fractures."

The medical profession, including the USFDA, should have borne these second thoughts in mind during the subsequent 25 years of expansion of the use of these drugs beyond any estimate that would have been made by these two scientists in the last decade of the 20th century.

Conclusion

There would be reason enough to feel confident that the legacy diseases associated with the phosphorus match industry were due to chronic exposure to smoke, based solely upon the contemporary evidence that the greater the exposure to the smoke, the greater the incidence of the effects, satisfying the important dose/effect relationship in considering possible causation (Bradford-Hill 1965). That this smoke consisted of a high concentration of pyrophosphoric acid (24%) and mixed P3 to P14 polyacids (52%) with the balance being almost entirely the relatively innocuous orthophosphate has been discussed, as well as why the only other likely candidates, elemental white phosphorus and the phosphorus oxides, can be ruled out. We have shown (Hinshaw and DeLong 2020) that in an experimental measurement (Fleisch et al. 1966), the increased accumulated mass of living chick femurs in tissue culture, properly interpreted in the light of understanding that cell-mediated resorption (resulting in mineral loss) is inhibited at low concentrations by either set of active agents in the smoke, PPi alone or Graham salt, a mixture of the lower polyphosphoric acids.

A simple process of elimination supports the etiology proposed: the smoke was the only suspect left after the investigations, conducted at the time of the legacy side effects, and other competing theories of causation were all rejected by experimental methods; the smoke, after many purely conjectural guesses over the years, has been found to consist only of the polyphosphoric acids, including PPi, and orthophosphate.

But the remarkable specificity of the same two major side-effects found to be associated with the smoke and the bisphosphonate drugs is placed beyond the possibility of mere coincidence due to the historical fact that these drugs, developed on the model of PPi, share to different degrees the antiresorptive character of PPi, and were pursued to the exclusion of continuing interest in PPi by Fleisch and his contemporary workers, partially due to the improper laboratory techniques used to assess the biological activity of PPi leading to the conclusion that its activity is limited in vivo by rapid hydrolysis. This latter error was rapidly corrected in the laboratories of Orimo and Rasmussen as is detailed in the previous chapter (Hinshaw and DeLong, 2020).

There is a physical model available for the interaction between the bisphosphonates and human bone (Mukerjee, Song, and Oldfield 2008). Using ^{2}H, ^{13}C, ^{15}N, and ^{31}P nuclear magnetic resonance spectroscopy, it has been demonstrated that several of the so-called "nitrogen containing bisphosphonates" bind to human hydroxyapatite in a manner in which the bisphosphonate group displaces an orthophosphate group from the bone mineral, thus "chemically" binding to the bone in a fixed configuration. The side chain, however, can jump from one ionized acid group to another. The binding constants vary widely with the side chain, as has been demonstrated with the differential rates of elution of mixtures of these compounds from hydroxyapatite columns with phosphoric acid (Nancollas et al. 2006). The rapid binding of pyrophosphoric acid to hydroxyapatite was established in the early modern studies (Jung et al. 1970; Bisaz, Jung, and Fleisch 1973 and 1978), and certified by a recent review (Orriss, Arnett, and Russell 2013). The similarity to the Mukerjee, Song, and Oldfield findings included the demonstration of displacement of orthophosphate by PPi. Interestingly, the binding of PPi was stronger than that of two non-nitrogen bisphosphonates comparators. Lacking the nitrogenous side chains apparently eliminates an additional binding enjoyed by the "nitrogen containing bisphosphonates" tested by Mukerjee, Song, and Oldfield.

An animal model is also desirable to test this hypothesis on an experimental basis in addition to the precedence and circumstantial logic employed here. Furthermore, the accumulation of these and other facts supports the possibility that an animal model to test this concept should be easy to design and inexpensive to conduct. But are there adequate animal models, in which pyrophosphate could be substituted for the bisphosphonate, of these two side-effects associated with the bisphosphonate drugs? The earliest successful model (Gotcher and Jee 1981) associating a bisphosphonate with the development of a syndrome quite similar to human ONJ was accomplished in a rat bred for susceptibility to periodontal disease. There are several mouse models for ONJ including an extremely well-documented ligature-induced periodontitis (de Molon et al. 2015). With respect to the AFFs, the availability of models is more limited. Very few non-human species are bipedal upright ambulators, and none of them are acceptable as experimental substitutes. The positional associated stress factors have been recognized as significant contributors to the location of this lesion (Koch 1917). However, many effects of bisphosphonate drug chronic administration on the tissue composition and/or the mechanical properties of bone have been demonstrated. Such test using chronic administration of PPi would easily be adaptable to following the established procedures (examples: Allen, Kubek and Burr 2010, Smith and Allen, 2013).

To recapitulate the possibilities for a model:

PPi and the polyacids are the only significant volumetric exposure in the legacy cases, and thus PPi alone would be satisfactory to compare with controls in a model (Spanggord 1985).

PPi has been demonstrated to be transferred to the blood through inhalation, although the quantitative measure of the degree of this transfer was hampered by lack of understanding of the rapid transfer of the compound from blood to a second compartment, evidently bone.

Thus a parenteral route of administration, preferably intravenous, would be sufficient to mimic the legacy circumstances (Isitman et al. 1988).

PPi is rapidly taken up by bone and is thus analogous to the first stage of the accepted mechanism of action of the bisphosphonate drugs (Jung et al. 1970, Orriss et al. 2016).

PPi has antiresorptive properties parallel to the bisphosphonate drugs, possibly but not necessarily low-dose dependent. A dose-response model design would be important (Fleisch et al. 1966, Orimo et al. 1969, Rasmussen et al., Delong et al. 1971).

Chronic exposure to PPi or the bisphosphonate drugs are both associated with the same pair of unusual side effects, and evaluation of substitutes for these factors would be the desirable end-points of such a model (Marx 2003, Odvina et al. 2003, Khosla et al. 2007, Shane et al. 2010 and 2013).

Finally, there remains one additional clue to suggest a possibly successful simple animal model. At the end of Rasmussen's definitive paper (Delong et al. 1971) on the inhibition of cell-mediated resorption in the rat, he mentions that there was an inexplicable persistent increase in the urinary phosphate excretion in animals pretreated with a prolonged exposure to intravenous infusion of both PPi and PTH for 16 hours followed by PTH alone for an additional six hours. After 16 hours of infusion of PTH alone, the serum phosphate level remained unaffected, whereas it is significantly elevated at the end of a simultaneous infusion of PTH and PPi for 16 hours. But if the PTH infusion is continued for another 6 hours after the PPi has been stopped, the serum phosphate remains elevated as if the PPi was still effective. The serum calcium was also elevated under these conditions. He speculated that this may have been due to a cellular effect of the long exposure to PTH. Should something like this be happening, it suggests that chronic treatment of the rat with PPi might produce a permanent change in the composition of the bone. This is mentioned only to inform the reader of the thoughts of Rasmussen. The effect could also be an indication of unanticipated persistence of the PPi in bone, an effect that would not necessarily have been noticed in the absence of the counter-stimulation of the PTH.

REFERENCES

Allen, Matthew R., Daniel J. Kubek, and David B. Burr. 2010. "Cancer Treatment Dosing Regimens of Zoledronic Acid Result in Near-Complete Suppression of Mandible Intracortical Bone Remodeling in Beagle Dogs." *Journal of Bone and Mineral Research* 25: 98–105.

Bisaz, Sylvia, A. Jung, and Herbert Fleisch. 1978. "Uptake by bone of pyrophosphate, diphosphonates and their technetium derivatives." *Clinical Science and Molecular Medicine* 54: 265-272.

Bradford Hill, Austin. 1965. "The Environment and Disease: Association or Causation?" *Proceedings of the Royal Society of Medicine, Section of Occupational Medicine* 58: 295-300.

Corbridge, D. C. E. 1978. *Phosphorus*. Amsterdam: Elsevier.

Dearden, William F. 1899. "Fragilitas ossium amongst workers in Lucifer match factories." *British Medical Journal* 270-271.

Dearden, William F. 1901. "The Causation of Phosphorus Necrosis." *British Medical Journal* 408-410.

Delong, Allyn, J. Feinblatt, and H. Rasmussen, 1971. "The effect of pyrophosphate infusion on the response of the thyroparathyroidectomised rat to parathyroid hormone and adenosine-3,' 5'-cyclic monophosphate." *Calcified Tissue Research* 8: 87–95.

De Molon, Rafael Scaf, Hiroaki Shimamoto, Olga Bezouglaia, Flavia Q. Pirih, Sarah M. Dry, Paul Kostenuik, Rogely W. Boyce, Denise Dwyer, Tara L. Aghaloo, and Sotirios Tetradis. 2015. "OPG-Fc but Not Zoledronic Acid Discontinuation Reverses Osteonecrosis of the Jaws (ONJ) in Mice." *Journal of Bone and Mineral Research* 30: 1627-40. doi: 10.1002/jbmr.2490.

Edwards, Beatrice J., John W. Hellstein, Peter L. Jacobson, Steven Kaltman, Angelo Mariotti, and Cesar A. Migliorati. 2006. "Expert Panel Recommendations: Dental Management of Patients on Oral Bisphosphonate Therapy." *American Dental Association Report of the Council on Scientific Affairs.*

Emsley, John. 2000. *The 13th Element.* New York: John Wiley & Sons.

Fleisch, Herbert, Fritz Straumann, Robert Schenk, Sylvia Bisaz, and Martin Allgöwer. 1966. "Effect of condensed phosphates on calcification of chick embryo femurs in tissue culture." *American Journal of Physiology* 211: 821.

Fleisch, Herbert, R. Graham G. Russell, Sylvia Bisaz, P. A. Casey, and R. C. Mühlbower. 1968. "The Influence of Pyrophosphate Analogues (Diphosphonates) on the Precipitation and Dissolution of Calcium Phosphate in Vitro and in Vivo." *Calcified Tissue Research* 2: Supp.

Fleisch, H. 1998. "Bisphosphonates in the treatment of osteoporosis." *Scandinavian Journal of Rheumatology* 27 Supplement 107:64.

Fleisch, Herbert. 1998b. "Bisphosphonates: Mechanisms of Action." *Endocrine Reviews* 19: 80–100.

Fraser, Donald. 1957. "Hypophosphatasia." *American Journal of Medicine* 22: 730-746.

Gotcher, J. E. and W. S. S. Jee. 1981. "The progress of the periodontal syndrome in the rice rat. II. The effects of a diphosphonate on the periodontium." *Journal of Periodontal Research* 16: 441-455.

Grasko J. M., R. P. Herrman, and S. D. Vasikaran. 2009. "Recurrent low-energy femoral shaft fractures and osteonecrosis of the jaw in a case of multiple myeloma treated with bisphosphonates." *Journal of Oral & Maxillofacial Surgery.* 67: 645-9.

Hinshaw, William Banks. 2007. "Bisphosphonates and Inorganic Phosphorus Chemistry." Abstract P43, *American Society of Bone and Mineral Research Topical Meeting on Bone Remodeling,* Washington.

Hinshaw, William Banks, and Louis DuBose Quin. 2013. "Using Medicinal Chemistry to Solve an Old Medical Mystery." *ACS Medicinal Chemistry Letters* 4: 2-4. doi: 10.1021/ml300430j.

Hinshaw, William Banks, and Louis Dubose Quin. 2015. "Recognition of the Causative Agent of "Phossy Jaw" and "Fragile Femur" in Fumes Arising from White Phosphorus." *Phosphorus, Sulfur, and Silicon* 190: 2082–2093.

Hinshaw, William Banks, and Allyn F. DeLong. 2016, "An Evaluative History of Bisphosphonate Drugs: Dual Physiologic Effects of Pyrophosphate as Inspiration for a Novel Pharmaceutical Class. *Journal of Osteoporosis*, 2016: 1-7. doi: 10.1155/2016/1426279.

Hinshaw, W. Banks, and Jennifer P. Schneider. 2018. "The Atraumatic Metatarsal Fracture: A Clinical Sign of Bisphosphonate-Associated Bone Fragility." *Journal of Advances in Medical and Pharmaceutical Sciences* 17: 1-7. doi: 10.9734/JAMPS/2018/ 42283.

Hinshaw, William Banks, and Allyn F. Delong. 2020. "The Pharmacology of Pyrophosphoric Acid-- A Critical Analysis of the First Quarter Century of Research." in *A Closer Look at Pyrophosphphates*. New York: NOVA Publications.

Isitman, Ali T., B. David Collier, David W. Palmer, LisaAnn Trembath, Arthur Z. Krasnow, Shyam A. Rao, Robert S. Hellman, Raymond G. Hoffman, David C. Peck, and Charles J. Dellis. 1988. "Comparison of Technetium-99m Pyrophosphate and Technetium-99m DTPA Aerosols for SPECT Ventilation Lung Imaging." *Journal of Nuclear Medicine* 29:1761-1767.

Jung A., R. G. G. Russell, S. Bisaz, D. B. Morgan, and H. Fleisch. 1970. "Fate of intravenously injected pyrophosphate-^{32}P in dogs." *American Journal of Physiology*. 218: 1757-1764.

Jung et al., R. G. G. Russell, S. Bisaz, A. Donath, D. B. Morgan, and H. Fleisch. 1971. "Inorganic Pyrophosphate in Plasma in Normal Persons and in Patients with Hypophosphatasia, Osteogenesis Imperfecta, and Other Disorders of Bone." *The Journal of Clinical Investigation* 50: 961-969.

Jung A., S. Bisaz, and H. Fleisch. 1973. "The Binding of Pyrophosphate and Two Diphosphonates by Hydroxyapatite Crystals." *Calcified Tissue Research* 11: 269-280.

King, W. R. and J. E. Peterson. 1978. "Phosphorus vapor exposure in phosphorus using plants." *American Industrial Hygiene Journal* 39: 922-7.

Khosla, Sundeep, David Burr, Jane Cauley, David W. Dempster, Peter R. Ebeling, Dieter Felsenberg, Robert F. Gagel, Vincente Gilsanz, and many others. 2007. "Bisphosphonate-Associated Osteonecrosis of the Jaw: Report of a Task Force of the American Society for Bone and Mineral Research." *Journal of Bone and Mineral Research* 22: 1479-1491.

Koch J. C. 1917. "The laws of bone architecture." *American Journal of Anatomy* 2: 242-243.

Kukher, V. T. and H. R. Hudson. 2000. *Aminophosphonic and Aminophosphinic Acids*. John Wiley & Sons: Chichester. See Chapter 1.

Lam, David K., George K. B. Sándor, Howard I. Holmes, Robert P. Carmichael, and Cameron N. K. Klokie. 2007. "Marble Bone Disease: A Review of Osteopetrosis and Its Oral Health Implications for Dentists." *Journal of the Canadian Dental Association* 73: 839-843.

Manolagas, Stavros C., Juliet Compston, Sudhaker Rao, and Ego Seeman. 2015. "A. Michael Parfitt: Enlightened Scholar and Revered Mentor May 10, 1930–May 18, 2015." *Journal of Bone and Mineral Research* 30: 1349–1355. doi: 10.1002/jbmr.2576.

Marx, Robert E. 2003. "Pamidronate (Aredia) and zoledronate (Zometa) induced avascular necrosis of the jaws: a growing epidemic." *Journal of Oral and Maxillofacial Surgery* 61: 1115-1118.

Marx, Robert E. 2008. "Uncovering the Cause of 'Phossy Jaw" Circa 1858 to 1906. Oral and Maxillofacial Surgery Closed Case Files:

Case Closed." *Journal of Oral and Maxillofacial Surgery* 66: 2356-2363.

Mukerjee, Sujoy, Yongcheng Song, and Eric Oldfield. 2008 "NMR Investigations of the Static and Dynamic Structures of Bisphosphonates on Human Bone: a Molecular Model." *Journal of the American Chemical Society* 130: 1264-1273.

Nancollas, G. H., R. Tang, R. J. Phipps, Z. Henneman, S. Gulde, W. Wu, A. Mangood, R. G. G. Russell, F. H. Ebitino. 2006. "Novel insights into actions of bisphosphonates on bone: Differences in interactions with hydroxyapatite." *Bone* 38: 617-627.

Odvina, C. V, D. S. Rao, C. Y. C. Pak, N. Maalouf, J. E. Zerwekh. 2003. "Adynamic Bone Disease during Bisphosphonate Therapy: Should We Be concerned?" Presentation Number: SU344. *Annual Meeting of the American Society of Bone and Mineral Research*.

Oliver, Thomas. 1902. *Dangerous Trades*. London: John Murray.

Orriss, Isabel R., Tim R. Arnett, and R. Graham G. Russell. 2016. "Pyrophosphate: a key inhibitor of mineralisation." *Current Opinion in Pharmacology* 28: 57–68.

Parfitt, A. M. 1996. "Skeletal heterogeneity and the purposes of bone remodeling: Implications for the understanding of osteoporosis." *Osteoporosis 1st ed*. San Diego: Academic Press. Page 326.

Pollock, Richard A., David M. Rubin, and Thomas Brown. 2015. ""Phossy Jaw" and "Bis-phossy Jaw" of the 19th and the 21st Centuries: The Diuturnity of John Walker and the Friction Match." *Craniomaxillofacial Trauma Reconstruction* 8: 262–270. doi: 10.1055/s-0035-1558452

Purcell, P. M., and I. W. Boyd. 2005. "Bisphosphonates and osteonecrosis of the jaw." *Medical Journal of Australia* 182: 417-418.

Rasmussen, H., J. Feinblatt, N. Nagata, Allyn DeLong. 1970. "Regulation of bone cell function." in *Osteoporosis* Barzel, U.S., ed. 187-198. New York-London: Grune & Stratton.

Rathbun, J. C. 1948. "Hypophosphatasia, a new developmental anomaly." *American Journal of Diseases of Children* 75: 822-31.

Reynolds, John J., Helen Murphy, R. C. Mühlbower, D. B. Morgan, and H. Fleisch. 1972." Inhibition by Diphosphonates of Bone Resorption in Mice and Comparison with Grey-Lethal Osteopetrosis." *Calcified Tissue Research* 12: 59-71.

Ruggiero S. L., Mehrotra B., Rosenberg T. J., Engroff S. L. 2004. Osteonecrosis of the jaws associated with the use of bisphosphonates: a review of 63 cases. *Journal of Oral and Maxillofacial Surgery* 62: 527-534.

Ruggiero S. L., J. Gralow, Robert E. Marx, A. O. Hoff, M. M. Schubert, J. M. Huryn, B. Toth, K. Damato, and V. Valero. 2006. Practical Guidelines for the Prevention, Diagnosis, and Treatment of Osteonecrosis of the Jaw in Patients with Cancer. *Journal of Oncology Practice* 2: 7-14.

Schenk, R., W. A. Mertz, R. Mühlbauer, R. G. G. Russell, and H. Fleisch. 1973. "Effect of Ethane-1-Hydroxy-1,1-Diphosphonate (EHDP) and Dichloromethylene Diphosphonate (Cl2MDP) on the Calcification and Resorption of Cartilage and Bone in the Tibial Epiphysis and Metaphysis of Rats." *Calcified Tissue Research* 11: 196-214.

Shane, Elizabeth, David Burr, Peter R. Ebeling, Bo Abrahamsen, Robert A. Adler, Thomas D. Brown, Angela M. Cheung, Felicia Cosman, Jeffrey R. Curtis, Richard Dell, David Dempster, Thomas A. Einhorn, Harry K. Genant, Piet Geusens, Klaus Klaushofer, Kenneth Koval, Joseph M. Lane, Fergus McKiernan, Ross McKinney, Alvin Ng, Jeri Nieves, Regis O'Keefe, Socrates Papapoulos, Howe Tet Sen, Marjolein CH van der Meulen, Robert S. Weinstein, and Michael Whyte. 2010. "Atypical Subtrochanteric and Diaphyseal Femoral Fractures: Report of a Task Force of the American Society for Bone and Mineral Research." *Journal of Bone and Mineral Research* 25: 2267–2294. doi: 10.1002/jbmr.253.

Shane, Elizabeth, David Burr, Bo Abrahamsen, Robert A. Adler, Thomas D. Brown, Angela M. Cheung, Felicia Cosman, Jeffrey R. Curtis,

Richard Dell, David W. Dempster, Peter R. Ebeling, Thomas A. Einhorn, Harry K Genant, Piet Geusens, Klaus Klaushofer, Joseph M Lane, Fergus McKiernan, Ross McKinney, Alvin Ng, Jeri Nieves, Regis O'Keefe, Socrates Papapoulos, Tet Sen Howe, Marjolein CH van der Meulen, Robert S Weinstein, and Michael P Whyte. 2013. "Atypical Subtrochanteric and Diaphyseal Femoral Fractures: Second Report of a Task Force of the American Society for Bone and Mineral Research." *Journal of Bone and Mineral Research* 29: 1–23.

Smith, Erik R. and Matthew R. Allen. 2013. "Bisphosphonate-induced reductions in rat femoral bone energy absorption and toughness are testing rate-dependent." *Journal of Orthopaedics* 31: 1317–1322. doi: 10.1002/jor.22343.

Spanggord, Robert J., Robert Rewick, Chou Tsong-Wen, Robert Wilson, R. Thomas Podoll, Theodore Mill, Richard Parnas. Robert Platz, and Daryl Roberts. 1985. "Environmental Fate of White Phosphorus/Felt and Red Phosphorus/Butyl Rubber Military Screening Smokes." *U. S. Army Medical Research and Development Command*, Fort Detrick, Frederick, Maryland 21701-5012.

Spiegel A. 2009. *"How A Bone Disease Grew To Fit The Prescription,"* All Things Considered (National Public Radio), December 21, 2009. Online: http://www.npr.org/2009/12/21/121609815/how-a-bone-disease-grew-to-fit-the-prescription.

Sutton, Roger A. L., Steven Mumm, Stephen P. Coburn, Karen L. Ericson, and Michael P. Whyte. 2012. "Atypical Femoral Fractures' during Bisphosphonate Exposure in Adult Hypophosphatasia." *Journal of Bone and Mineral Research* 27: 987–994.

Thorpe, T. E., Thomas Oliver, and George Cunningham. 1898. *Reports on the Use of Phosphorus in the Manufacture of Lucifer Matches.* London: Eyre & Spottiswoode.

U.S. Department of Health and Human Services Food and Drug Administration. 2005. *"Reviewer Guidance Evaluating the Risks of Drug Exposure in Human Pregnancies."* April.

Van Wazer, J. R. 1966. *Phosphorus and Its Compounds*. New York: Interscience.

Von Steubenrauch, L. 1899. "Experimentelle Untersuchungen uber Phosphornekrose [Experimental studies on phosphorus necrosis]." *Arkiv für klinische chirurgie* 59: 144-152.

Von Steubenrauch, L. 1900. "Ueber die unter dem Einflusse des Phosphors entstahenden Veränderungen des wachsenden Knochens [About the changes in the growing bone that occur under the influence of phosphorus]." *Sitzungsberichte der Gesellschaft für Morphologie und Physiologie in Munchen* 16:3-35.

Whyte, Michael P., Deborah Wenkert, Karen L. Clements, William H. McAlister, and Steven Mumm. 2003. "Bisphosphonate-Induced Osteopetrosis." *New England Journal of Medicine* 349: 457-63.

Whyte, Michael P. 2009. "Atypical Femoral Fractures, Bisphosphonates, and Adult Hypophosphatasia." *Journal of Bone and Mineral Research* 24: 1131-1134.

Yates, Christopher J., Miriam J. Bartlett, and Peter R. Ebeling. 2010. "An Atypical Subtrochanteric Femoral Fracture from Pycnodysostosis - A Lesson from Nature." *Journal of Bone and Mineral Research* 26: 1377–1379. doi: 10.1002/jbmr.308.

ABOUT THE EDITORS

Dr. Ritesh L. Kohale
Assistant Professor
Department of Physics, Sant Gadge Maharaj Mahavidyalaya, Hingna
Nagpur, 44110 (India)

Dr. Ritesh L. Kohale obtained his M.Sc. degree in Physics from RTM Nagpur University, Nagpur, India in 2008. He obtained his Ph.D. in 2014 in Physics on Rare Activated Pyrophosphate Phosphors for Solid State Lighting from RTM Nagpur University, Nagpur. Dr. Ritesh L. Kohale is presently working as Assistant Professor and Head in Department of Physics, Sant Gadge Maharaj Mahavidyalaya, Hingna Nagpur, India and he is having 9 years of teaching experience. During his research career, he involved in the synthesis and characterization of solid state lighting materials, pyrophosphate phosphors. Dr. Kohale published more than 35 (thirty five) research papers in international and national reviewed journals and presented more than 10 research papers in international and national conferences on solid-state lighting, LEDs.

He is member of numerous professional organizations such as American Association of Physics Teachers (AAPT), Luminescence Society of India (LSI), and International Association of Engineers

(IAENG). Dr. Kohale is an editorial board member of American Journal of Optics and Photonics He is reviewer of international journals viz. ELSEVIER Publication, USA and John Wiley & Sons Inc publication, USA. Apart from this he also carried out his research on expansion of universe, Isoredshift and Blueshift and published numerous research articles. He visited abroad the countries viz. Greece, China and Hong-Kong for research and education purpose.

Dr.Sanjay J.Dhoble
Professor
Department of Physics,R.T.M.Nagpur University, Nagpur-440033, India

Prof. Sanjay J. Dhoble, obtained M.Sc. degree in Physics from Rani Durgavati University, Jabalpur, India in 1988. He obtained his Ph.D. degree in 1992 on Solid State Physics from Nagpur University, Nagpur. Dr. S.J. Dhoble is presently working as Professor in Department of Physics, R.T.M. Nagpur University, Nagpur, India. During his research career, he is involved in the synthesis and characterization of solid state lighting materials, EVI parameter as well as development of radiation dosimetry phosphors using thermoluminescence, mechanoluminescence and lyoluminescence techniques and utilization of fly ash. Dr. Dhoble published more than six hundred thirty four research papers in international and national reviewed journals and presented more than 1346 research papers in international and national conferences on solid-state lighting, LEDs, radiation dosimetry, phototherapy and laser materials. As per SCOPUS database, he holds the top position in the world in terms of publications on phosphor, lamp phosphor, radiation dosimetry materials. He is elected as the Fellow of Luminescence Society of India. He recently received Career-360 Best Faculty Research Awards-2018, Best Research Award-2016, R.T.M. Nagpur University,

About the Editors

Prof. B.T.Deshmukh Research Award 2016, Prof. B.P.Chandra Research Award 2016, Outstanding Scientist-2015 award by Venus International Foundation, Advanced Materials Scientist Letter Awards-2011 by VBRI Press. He is co-author of 24 books/Chapters, three of which are of international level published by Springer Science, Heidelberg, New York, (Series: Springer Series in Materials Science), CRC Press, Taylor and Francis Group, Trans Tech Publication, Switzerland, Nova Science, New York, Lambert Academic Publishing respectively and one of book published on OLED by Elsevier Imprint Woodhead Publication, USA. He is life member of Society for Materials Chemistry, Luminescence Society of India, Indian Association for Radiation protection, Indian Laser Association, Indian Physics Association, International Association of Advanced Materials (IAAM), Vidarbha Environment Society, Nuclear Track Society of India and Indian Physics Teachers' Association, Society for Technologically Advanced Material of India (STAMI), Indian Science Congress

INDEX

A

acid, 1, 2, 5, 7, 8, 11, 13, 29, 30, 31, 34, 48, 71, 72, 73, 74, 77, 78, 84, 90, 91, 95, 96, 97, 102, 103, 108, 109
alkaline phosphatase, 69, 74, 105
ammonium, 2, 26, 31, 51, 52
annealing, 33, 36, 37, 62, 66
applications, vii, viii, 2, 4, 12, 16, 21, 22, 24, 39, 43, 48, 49, 51, 60, 61, 63, 67
arthritis, 18, 63, 65
atmosphere, 15, 33, 34
atoms, 2, 5, 22, 48, 97
ATP, 12, 13, 14

B

batteries, 21, 22, 40, 44
behaviors, 22, 23, 63
biochemistry, 1, 22, 95
blood, 13, 51, 79, 99, 105, 110
bonding, 5, 11, 12
bone, 12, 18, 51, 63, 65, 71, 72, 73, 75, 76, 77, 78, 79, 80, 81, 82, 83, 84, 86, 87, 90, 91, 92, 93, 95, 97, 99, 100, 101, 102, 106, 107, 109, 110, 111, 112, 115, 116, 118, 119
bone mass, 76, 106, 107
bone resorption, 81, 83, 86, 87, 90, 101, 107
bones, 4, 90, 91, 92, 101, 107

C

calcification, 13, 18, 19, 65, 75, 85, 113
calcination temperature, 27, 29, 30, 32
calcitonin, 17, 81, 82
calcium, 12, 19, 31, 44, 51, 55, 65, 67, 68, 73, 74, 76, 77, 79, 80, 81, 82, 83, 85, 100, 111
candidates, 16, 22, 108
cathode materials, 22, 40, 49, 51
cation, 6, 9, 10, 22, 23
causation, 95, 103, 104, 108
ceramic, 41, 42, 43, 65
chemical, viii, 1, 4, 15, 21, 22, 23, 24, 26, 27, 29, 31, 36, 37, 38, 40, 51, 52, 71, 76, 80, 95, 100
color, 23, 25, 38, 44, 50, 52, 57, 61, 67
combustion, 23, 29, 33, 37, 38, 42, 56, 67, 93

composition, 8, 11, 110, 111
compounds, 4, 8, 9, 16, 23, 48, 50, 51, 75, 76, 78, 83, 87, 100, 102, 109
crystal structure, 5, 8, 23, 24, 25, 30, 37, 41, 64
crystalline, 3, 7, 22, 23, 34, 49, 62, 66, 67, 78, 79
crystallinity, 21, 37, 58
crystallization, 36, 50, 64, 74, 78
culture, 74, 85, 86, 87, 108, 113

D

deposition, 19, 36, 44, 75, 76, 79
derivatives, 1, 7, 10, 112
diffusion, 4, 16, 51, 52, 68
diseases, 84, 93, 95, 108
distilled water, 29, 34, 52
DNA, 12, 13, 51, 63
dogs, 79, 86, 99, 100, 114
doping, 23, 42, 44, 50, 51, 60, 66
drugs, 84, 90, 91, 92, 102, 103, 104, 107, 108, 109, 110, 111
drying, 31, 32, 34

E

emission, 2, 15, 17, 23, 37, 38, 40, 42, 44, 49, 50, 53, 55, 56, 57, 58, 59, 61, 62, 66, 68
energy, viii, 1, 12, 14, 22, 27, 38, 40, 48, 49, 50, 51, 53, 54, 56, 57, 58, 59, 61, 63, 67, 113, 118
energy transfer, 38, 48, 50, 53, 54, 56, 57, 58, 59, 61, 63, 67
environment, viii, 17, 73, 112, 123
enzyme, 12, 74, 105
etiology, 91, 103, 108
europium, 15, 32, 65
evidence, 7, 73, 75, 84, 90, 91, 94, 95, 100, 102, 104, 108

excitation, 38, 44, 49, 50, 53, 55, 58, 59, 60, 62, 66
excretion, 82, 83, 111
exposure, 84, 93, 99, 100, 102, 106, 108, 110, 111, 115

F

factories, 91, 100, 112
femur, 76, 80, 81, 83, 90, 100, 102, 104
force, 90, 103, 104
formation, 8, 28, 31, 73
formula, 2, 22, 23, 31, 61
fractures, 90, 92, 100, 101, 102, 104, 105, 106, 107, 113
fragility, 90, 92, 101, 107

G

gel, 30, 31, 32, 34, 37, 38, 42
geometry, 5, 11, 12
globalization, viii
growth, 76, 79, 107

H

host, 3, 42, 49, 51, 54, 60, 66
human, 19, 72, 73, 92, 95, 99, 101, 109, 110
hydrogen, 2, 6, 8, 32, 51, 77
hydrolysis, 9, 12, 13, 31, 73, 80, 82, 83, 96, 97, 99, 102, 105, 109
hydroxyapatite, 13, 80, 86, 87, 109, 116

I

in vitro, 74, 75, 87
in vivo, 73, 102, 109
incidence, 95, 99, 103, 108
India, 1, 21, 47, 121, 122
individuals, 90, 92, 105

industries, 3, 72, 90
industry, 1, 3, 49, 51, 62, 63, 89, 92, 94, 95, 99, 100, 106, 108
inhibition, 76, 77, 78, 79, 82, 86, 101, 102, 105, 107, 111
inhibitor, 13, 51, 63, 73, 74, 79, 80, 83, 85, 86, 116
intravenously, 73, 79, 80, 86, 103, 114
ions, 2, 3, 12, 15, 23, 38, 42, 48, 50, 52, 53, 54, 55, 56, 57, 58, 59, 60, 62, 63, 64, 68

L

lanthanide, 2, 9, 47, 50, 51, 59, 63, 66
lanthanum, 43, 44, 66
lead, 76, 82, 90, 107
lesions, 93, 99, 100, 101
light, 2, 17, 21, 22, 49, 53, 61, 62, 66, 67, 68, 99, 106, 108
light emitting diode, 21, 49, 61, 68
luminescence, 42, 44, 64, 65, 67, 68

M

magnesium, 31, 32, 41, 85
mass, 2, 75, 77, 78, 79, 96, 107, 108
materials, vii, 2, 3, 4, 7, 9, 15, 21, 22, 23, 24, 25, 26, 27, 28, 29, 31, 32, 33, 34, 36, 37, 39, 40, 47, 49, 50, 51, 52, 56, 61, 63, 121, 122
measurement, 56, 75, 81, 107, 108
measurements, 27, 29, 52
media, 4, 13, 14, 28
medical, 77, 84, 91, 92, 100, 103, 107, 108
medicine, 72, 99, 103
metabolism, 17, 73, 81, 85, 90, 107
metal ion, 50, 60, 63
mice, 76, 77, 79, 81, 101
mineralization, 73, 75, 79, 84, 102, 105
mixing, 26, 27, 28, 29, 32
models, 79, 91, 110

molecules, 4, 6, 8, 49, 90
morphology, 22, 25, 37, 39, 41

N

nanoparticles, 27, 43, 68
nickel, 24, 40, 41

O

ONJ, 91, 92, 101, 102, 103, 105, 110, 112
optical properties, 10, 23, 41, 47, 52, 61, 63, 65, 67
osteonecrosis of the jaw, 90, 91, 113, 116
osteoporosis, 91, 105, 107, 113, 116
oxidation, 4, 6, 93, 96
oxygen, 2, 5, 8, 10, 11, 96, 97, 100

P

parallel, 8, 72, 79, 91, 102, 111
parathyroid, 17, 77, 85, 87, 112
parathyroid hormone, 17, 77, 85, 87, 112
permission, 57, 58, 76
pH, 31, 32, 34, 73, 77, 83, 99
phosphate(s), viii, 1, 2, 3, 4, 5, 6, 7, 8, 9, 10, 11, 12, 13, 15, 17, 19, 22, 26, 31, 32, 36, 41, 44, 48, 50, 51, 52, 64, 68, 74, 76, 80, 85, 87, 111, 113
phosphorus, 1, 2, 4, 5, 22, 48, 74, 75, 77, 84, 89, 90, 91, 93, 94, 95, 96, 99, 100, 106, 108, 115, 119
photoluminescence, 10, 15, 22, 25, 37, 39, 44, 48, 50, 51, 53, 55, 60, 61, 62, 63, 65, 66, 67, 68
physical properties, viii, 16, 48, 63, 64
polymerization, 13, 31, 32
polyphosphates, 13, 15, 17, 74, 75, 85
potassium, 2, 15, 60

precipitation, 29, 31, 32, 35, 38, 42, 43, 51, 52, 62, 68, 74
preparation, 25, 26, 27, 29, 34, 36, 68, 82
pyrophosphate(s), vii, viii, 1, 2, 10, 11, 12, 13, 14, 15, 16, 17, 19, 21, 22, 23, 24, 25, 26, 27, 28, 29, 31, 32, 33, 34, 35, 36, 37, 38, 39, 40, 41, 42, 47, 48, 49, 50, 51, 52, 53, 55, 59, 60, 61, 62, 63, 64, 65, 68, 69, 72, 73, 75, 85, 86, 87, 91, 110, 112, 114, 121

R

radiation, 49, 50, 55, 62, 63, 122, 123
rare-earth activated pyrophosphates, viii
raw materials, 2, 26, 27, 29, 31
reactions, 13, 32, 95
researchers, vii, 26, 48, 82, 93
response, 77, 85, 94, 111, 112
rings, 7, 9, 10
RNA, 13, 51, 63
room temperature, 24, 26, 29, 34, 96
routes, 22, 25, 39

S

salts, 4, 8, 10, 11, 50
science, vii, 1, 18, 42, 95
sensing, 39, 44, 45, 66, 69
serum, 74, 79, 81, 111
side effects, 84, 89, 90, 91, 102, 108, 111
skeleton, 77, 92, 101
sodium, 2, 10, 15, 40, 50, 64
sol-gel, 29, 31, 33, 37, 42
solid phase, 26, 29, 34
solid state, 23, 24, 26, 28, 38, 41, 49, 50, 51, 52, 55, 59, 60, 61, 62, 63, 68, 121, 122
solution, 7, 13, 23, 29, 30, 32, 33, 34, 37, 38, 48, 56, 81, 94, 107
species, 4, 6, 7, 8, 10, 16, 76, 110
speculation, 93, 95, 97

stability, 7, 15, 49, 65, 82
state(s), 1, 4, 5, 6, 7, 10, 11, 23, 26, 27, 37, 38, 40, 51, 53, 56, 65, 72, 107, 121, 122
steel, 24, 34, 52
stimulation, 77, 78, 112
strontium, 43, 52, 66
structure, 5, 7, 9, 11, 16, 17, 23, 24, 30, 40, 48, 64, 65, 78, 102
substrate, 24, 36, 105
symmetry, 5, 7, 9, 40, 64, 68
syndrome, 92, 101, 110, 113
synthesis, 10, 15, 21, 25, 26, 27, 28, 29, 31, 32, 33, 34, 35, 36, 37, 39, 42, 43, 47, 63, 64, 121, 122

T

Task Force, 104, 115, 117, 118
techniques, viii, 1, 22, 25, 34, 36, 37, 39, 75, 76, 91, 109, 122
teeth, 4, 90, 91
temperature, 2, 5, 15, 24, 26, 27, 28, 29, 30, 31, 32, 34, 36, 37, 39, 41, 44, 45, 48, 52, 60, 61, 65, 66, 96
therapy, 90, 93, 105, 106
thin films, 22, 24, 36, 41
tissue, 18, 65, 74, 75, 76, 85, 86, 87, 108, 110, 113
titanium, 44, 49, 64
transition metal, 11, 16, 51, 56, 59, 60, 63
transition metal ions, 16, 51, 59, 60, 63
treatment, 36, 44, 91, 105, 111, 113

U

urea, 29, 30, 42
urine, 13, 51, 74, 83, 85
USA, 43, 122, 123
UV light, 62

Index

V

victims, 93, 94, 95

W

water, 2, 6, 8, 15, 30, 31, 33, 34, 50, 74, 96, 97, 99
wavelengths, 53, 59, 60, 62
workers, 89, 90, 91, 92, 95, 96, 97, 100, 109, 112

X

XRD, 28, 29, 49, 62

Y

yield, 14, 39, 106